6th International Conference on Mechatronics and Mechanical Engineering (ICMME 2019)

IOP Conference Series: Materials Science and Engineering
Volume 717

Wuhan, China
9 - 11 November 2019

ISBN: 978-1-7138-2576-0
ISSN: 1757-8981

Printed from e-media with permission by:

Curran Associates, Inc.
57 Morehouse Lane
Red Hook, NY 12571

Some format issues inherent in the e-media version may also appear in this print version.

Institute of Physics
Dirac House, Temple Back
Bristol BS1 6BE UK

Phone: 44 1 17 929 7481
Fax: 44 1 17 920 0979

techtracking@iop.org

Additional copies of this publication are available from:

Curran Associates, Inc.
57 Morehouse Lane
Red Hook, NY 12571 USA
Phone: 845-758-0400
Fax: 845-758-2633
Email: curran@proceedings.com
Web: www.proceedings.com

TABLE OF CONTENTS

CONTROL THEORY AND SYSTEM

IOP Conf. Series: Materials Science and Engineering **717** (2020) 011001 doi:10.1088/1757-899X/717/1/011001

Preface

The 6th International Conference on Mechatronics and Mechanical Engineering (ICMME2019) was held at China University of Geosciences (Wuhan), China during November 9th-11th, 2019. The ICMME2019 (Wuhan) was co-sponsored by China University of Geosciences (Wuhan), China, Hefei University of Technology, China and Tianjin University, China.

The purpose of this meeting was to provide an international forum for recent developments in selected fields of research in mechatronics and mechanical engineering by bringing together researchers, practitioners and research scholars from all over the world to exchange and share their experiences and research results.

After several rounds of the review procedure, the program committee accepted papers for the publication in IOP Conference Series. We wish to express our sincere appreciation to all the individuals who have contributed to ICMME2019 (Wuhan). Special thanks are extended to the invited keynote and plenary speakers as well as all the authors for contributing their latest research to the conference.

This conference program was highlighted by four keynote and plenary speakers: Prof. Dennis K. Lieu, University of California, Berkeley, USA; Prof. Lianqing Yu, Wuhan Textile University, China; Prof. Mohd Hamdi Bin Abd Shukor, University of Malaya, Malaysia; Assoc. Prof. Suhana Mohd Said, University of Malaya, Malaysia. There were altogether four sessions during the conference. One best presentation was selected from each session. The best one was announced at the end of each session, and awarded the certificate. The conference was intended to provide chances for intensive discussions about their own selected topics, practical challenges encountered and identify future directions for the development of related modeling strategies and engineering techniques in the field of mechatronics and mechanical engineering.

We express our deepest gratitude to all members of the scientific committee, reviewers, sponsors, participants in the conference and members of the ICMME organizing committee for their valuable contribution to the successful organization of ICMME2019 (Wuhan).

Conference Committee Chair
Prof. Huafeng Ding,
China University of Geosciences, China

IOP Conf. Series: Materials Science and Engineering **717** (2020) 011002 doi:10.1088/1757-899X/717/1/011002

Committee

Conference Committee Chairs
Bin Zi, Hefei University of Technology, China
Haitao Liu, Tianjin University, China
Huafeng Ding, China University of Geosciences, China

Conference Program Committees
Chaolong Song, China University of Geosciences (Wuhan), China
Huijun Li, University of Wollongong, Australia

Conference Local Chairs
Jie Luo, China University of Geosciences (Wuhan), China
Tianye Huang, China University of Geosciences (Wuhan), China

Technical Committees
A. Y. Bani Hashim, Universiti Teknikal Malaysia Melaka, Malaysia
Arjon Turnip, Indonesian Institute of Sciences, Indonesia
FEI-LING Pua, Universiti Tenaga Nasional (UNITEN), Malaysia
G. Arunkumar, Sathyabama University, Chennai, India
Hendriko Hendriko, Politeknik Caltex Riau, Indonesia
HISAKI WATARI, Tokyo Denki University, Japan
M.P Jahan, Miami University, USA
Mohd Rizal Salleh, Universiti Teknikal Malaysia Melaka, Malaysia
N.Kanthavelkumaran, Ponjesly College of Engineering, Alamparai, Tamilnadu India
S. Rajesh, Sreenivasa Institute of Technology and Management Studies, India
Samir Mekid, King Fahd University of Petroleum and Minerals, Saudi Arabia
Somlak Wannarumon Kielarova, Naresuan University, Phitsanulok, Thailand
Suksan Prombanpong, King Mongkut's University of Technology Thonburi, Thailand
Tai-Shen Huang,Chaoyang University of Technology, Taiwan
Vu Ngoc Pi,Thai Nguyen University of Technology, Viet Nam
Yu-Che Huang, Chaoyang University of Technology, Taiwan

IOP Conf. Series: Materials Science and Engineering **717** (2020) 011003 doi:10.1088/1757-899X/717/1/011003

Peer review statement

All papers published in this volume of *IOP Conference Series: Materials Science and Engineering* have been peer reviewed through processes administered by the proceedings Editors. Reviews were conducted by expert referees to the professional and scientific standards expected of a proceedings journal published by IOP Publishing.

IOP Conf. Series: Materials Science and Engineering **717** (2020) 012001 doi:10.1088/1757-899X/717/1/012001

Fuel Properties, Performance and Emission of Alternative Fuel from Pyrolysis of Waste Plastics

W Khatha, S Ekarong, M Somkiat and S Jiraphon

Institute of Engineering, Suranaree University of Technology, Nakhon Ratchasima 30000 Thailand

vkata@sut.ac.th; ekarong@sut.ac.th; somkiat_mai@outlook.co.th; jiraphon@sut.ac.th

Abstract. Increase in energy demand, stringent emission norms and depletion of oil resources have led the researchers to find alternative fuels for internal combustion engines. Pyrolysis oil extracted from waste plastics, by pyrolysis, as a fuel for internal combustion engines has been demonstrated to be one of the best available waste management methods. This research aimed to study and compare fuel properties, engine performance and emission of alternative fuel from pyrolysis of waste plastics, which focused on physical and chemical properties of waste plastic crude oil (WPCO) and waste plastic oil (WPO) obtained by fractional distillation process. The result shows that the physical and chemical properties of WPCO are quite diffrent from commerical diesel while the fuel properties of WPO were found to be similar to that of diesel. The experimental results showed that the use of WPO led to an increase in brake specific fuel consumption but the slightly reduction in the brake thermal efficiency was founded with respect to diesel fuel. The combustion of WPO tended to increase NOX while a reduction in smoke was obtained compared with the pure diesel fuel.

1. Introduction

Plastic waste is a big issue in Thailand, because the amount of recycled plastic remains low due to the high investment requirements for recycling and the high operating costs of recycling facilities [1]. Plastic waste is an ideal source of energy due to its high heating value and abundance. Therefore, in recent years, various plastic to fuel (PTF) or waste to fuel (WTE) technologies such as gasification, pyrolysis, refuse derived fuel (RDF), and plasma arc gasification have gained significant attention for the management of plastic waste. The most attractive technique for chemical feedstock recycling is pyrolysis. Pyrolysis is a decomposition process of long-chain hydrocarbon (polymer) molecules into smaller sizes (monomer) with the use of high heat (450-800 ^{o}C), in a shorter duration and a condition with the absence of oxygen, generating products in form of carbon, as residues and volatile hydrocarbons which can be condensate as fuel and non-condensable as gaseous fuel. In addition, researchers showed that the properties of the liquid fuel from waste plastics are practically like diesel fuel and the engine could be operated using neat pyrolysis oil from waste plastics. Experimental engine results from previous studies have been reported that pyrolysis oil exhibited higher brake thermal efficiency and higher regulated exhaust emissions such as nitrogen oxides, hydrocarbon, and carbon monoxide with respect to diesel combustion [2]. In this study, it is aimed to study the characterization and fuel properties of alternative fuel from pyrolysis of waste plastics, which focused on physical and chemical properties of the fuel and compare engine performance and emissions.

2. Methodology

2.1. Materials
The raw materials utilized in this study were derived from waste plastics such as plastic bag waste collected from scavengers around Suranaree Subdistrict, Nakhon Ratchasima, Thailand. The compositions of the plastics were categorized as polyethylene (PE), and polystyrene (PS) of which approximately 70% were organic contaminated contents.

2.2. Conversion process
Pyrolysis process is the chemical decomposition of organic substances by heating in the absence of oxygen. The waste plastics is gently cracked by adding catalyst and the gases are condensed in a series of condensers to give a low sulphur content distillate. All this happens continuously to convert the waste plastics into fuel. All the gases from this process are treated before it is let out in atmosphere. The flue gas is treated through scrubbers and water/chemical treatment for neutralization. The non-condensable gas goes through water before it is used for burning. Since the plastics waste is processed about 300-350 ^{o}C and there is no oxygen in the processing reactor, most of the toxics are burnt.

2.3. Experimental setup
The engine used for this study is a four-stroke, single-cylinder, water-cooled, direct injection compression ignition engine. An eddy current dynamometer with a load cell was used to load the engine. The tests were performed at rated speed of 1500 rpm and three different engine loads (25%, 50% and 75% of maximum engine torque). Commercial diesel fuel was used as a baseline comparison with WPO for the engine run tests. An air box was used to measure the air flow rate to the engine and volumetric fuel flow rate was measured using a burette and stopwatch. The TESTO 350 analyzer was employed to measure nitrogen oxide (NOX) and TESTO 308 was used to evaluate smoke index. At every load, readings were taken after the engine reached steady state. For every refueling, engine was kept running for 10 minutes such that it consumes entire fuel entangled in fuel pipe lines. Repeatability of readings was guaranteed by duplicating the investigations thrice. The technical specifications of the engine are given in Table 1.

Table 1. Specifications of diesel engine.

Engine	Specifications
Model	Kirloskar TV1
Engine type	1 cylinder, 4 strokes,Water cooled, Direct injection
Bore x Stroke	87.5 x 110 mm
Swept volume	661 cc
Rate output	3.5 kW at 1500 rpm
Compression ratio	12:1-18:1
Dynamometer	Eddy current

3. Results

3.1. Chemical characterization
The pyrolysis oil or waste plastic crude oil (WPCO) and waste plastic oil (WPO) obtained by the fractional distillation used in this study is extracted from waste plastic. Gas chromatography mass spectrometry (GC-MS) was used to analyze the chemical compounds present in the WPCO compared with WPO. The GC-MS results of WPCO and WPO were presented in Figure 1. The major chemical compounds present in WPCO and WPO and their percentage area were listed in Table 2.

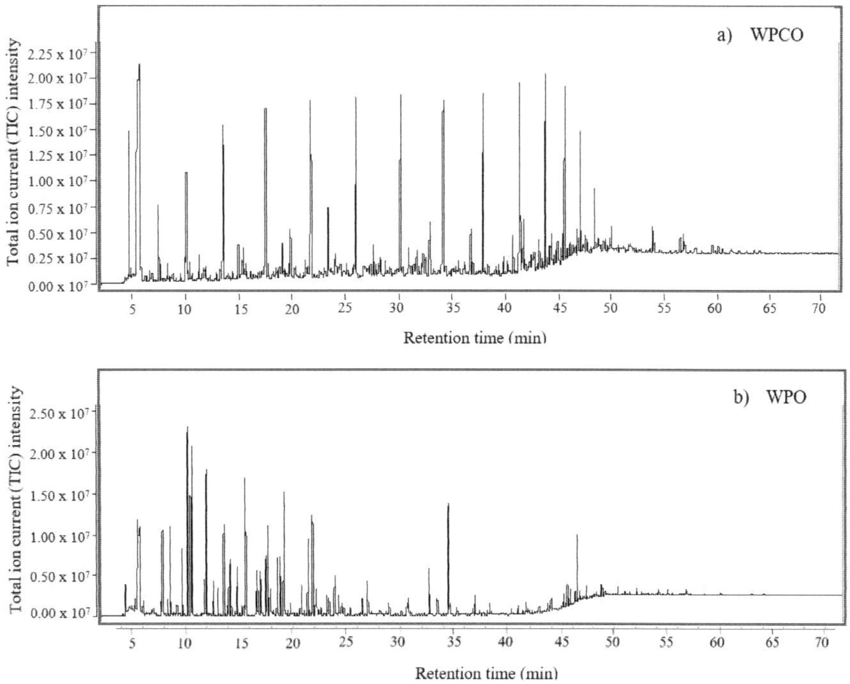

Figure 1. Total ion current chromatogram for WPCO and WPO.

Table 2. Principle constituents (area %) identified by GC-MS

Carbon content	WPCO	WPO
C4	2.77	Not detected
C5	0.84	Not detected
C6	3.76	Not detected
C7	5.38	Not detected
C8	34.02	1.03
C9	19.94	0.73
C10	28.49	2.28
C11	0.53	5.71
C12	0.75	9.78
C13	Not detected	10.08
C14	Not detected	11.03
C15	0.67	10.48
C16	0.19	8.71
C17	0.26	7.90
C18	0.32	8.88
C19	0.36	7.86
C20	0.24	8.10
C21	0.36	3.63
C22	0.27	2.29
C23	0.20	1.08
C24	0.19	0.43
C25	0.20	Not detected
C26	0.18	Not detected
C27	Not detected	Not detected
C28	0.08	Not detected

The WPCO and WPO contain many hydrocarbon compounds both light and heavy fractions and from the lowest carbon atom number (C4-C5) to the highest one (> C20). It could be classified into 3 groups to easily identify. C4-C11 group represents light or gasoline fraction due to the tendency that the largest percentage of carbon atom number in the gasoline is commonly around C7, C8, and C9. C11-C20 group could represent medium and heavy fractions, including diesel fraction. The diesel fuel fraction mostly contains high percentage of the atom number C16-C20 [3]. The result of this study showed that waste plastics composition strongly influences propensity formation of certain hydrocarbons group. The WPCO of this study leads to form mostly C12-C20 fraction. While the fractional distillation led to the trend of similarity between WPO and diesel fuel. It can be seen that chemical compounds were found to be closer to diesel fuel.

3.2. Physical and chemical properties

The basic physical and chemical properties of WPCO, WPO and commercial diesel fuel as reference fuels were measured according to ASTM standards and were shown in Table 3. The experimental results showed that WPCO possessed lower kinematic viscosity, flash point, distillation temperature, and gross calorific value while density and specific gravity were higher with respect to those of WPO. However, the fuel characteristics of WPO such as density, viscosity, and gross calorific value have similar properties with those of fossil fuels. Therefore, WPCO can't be used as alternative fuel for compression ignition engines due to the properties of WPCO were found to be closer to gasoline.

Table 3. Physical and chemical properties of test fuels.

Fuel property	Method (ASTM)	Diesel	CWPO	WPO
Density at 15.6 °C (kg/m^3)	D1298	834	899	823
Specific gravity at 15.6 °C	D4052	0.835	0.900	0.824
Kinematic viscosity at 40 °C (cSt)	D445	3.44	1.66	3.11
Flash point (°C)	D93	66	35	54
Distillation 90% vol. (°C)	D86	348	164	276
Calculate cetane index	D976	56.57	-	46.7
Gross calorific value (MJ/kg)	D240	45.56	37.72	45.24

3.3. Engine performance

In present section, performance parameters such as brake specific fuel consumption (BTE) and brake thermal efficiency (BTE) of the engine according to operation load for all the test fuels are presented and analyzed. The results showed in the plots correspond to an average of three different measurements. The variations of BSFC and BTE as a function of engine load for test fuels are shown in Figure 2. It can be seen that the BSFC decreased as the engine operating loads increased for all fuels. The lower BSFC implies that less amount of fuel is needed to generate unit power output. WPO presented higher BSFC than commercial diesel fuel due to its lower calorific value that required more quantity of oil to be burnt to produce the same power output similar to commercial diesel fuel (see Table 3) [4]. BTE indicates the ability of combustion system to accept the experimental fuel and provides a comparable means of assessing how efficiently the fuel was converted into mechanical output [5]. WPO showed a drop-in efficiency, since more energy is spent to breakdown the heavy hydrocarbon chains (C13 to C22) [4]. Besides, higher BSFC was obtained from the combustion of waste plastic oil, resulted in the lower BTE with respect to commercial diesel fuel for all engine loads tested.

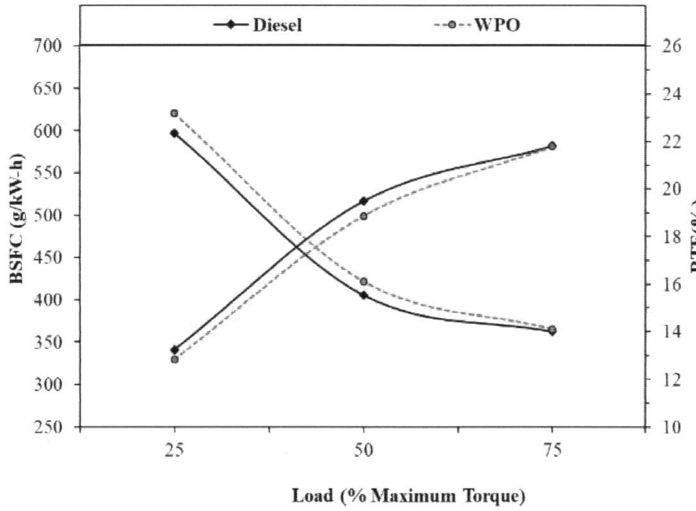

Figure 2. Variation of BSFC and BTE with load

3.4. Emissions

The emissions emitted from the engine, measured at tail pipe corresponding to variation of load are measured. The results showed in the plots correspond to an average of three different measurements. The variation of the NOX emissions and smoke opacity with engine loads and test fuels was shown in Figure 3.

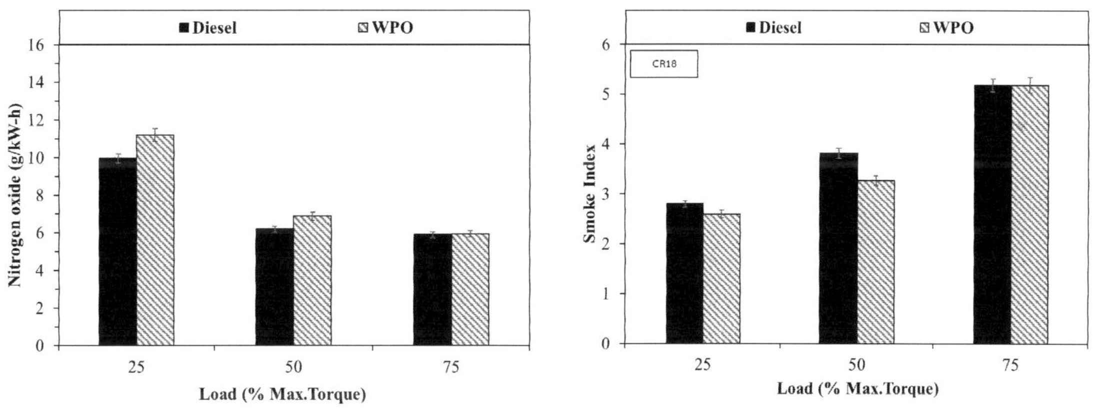

Figure 3. Variation of nitrogen oxide and smoke index with load.

Oxides of nitrogen present in emissions is the combination of nitric oxide (NO) and nitrogen dioxide (NO2). In diesel engines the mechanism that produces the majority of NOX is the thermal mechanism due to availability of oxygen, reaction time and combustion temperature [6,7]. It can be seen that the NOX emissions for WPO operation is higher than that for commercial diesel fuel. The reason that NOX emissions is the higher nitrogen content in the fuel that promotes the NOX formation by fuel mechanism. The variation of smoke opacity is attributed to the high overall equivalence ratio and the number of fuel rich regions in the combustion chamber when the engine is operated at high load, resulting in the high critical conditions for soot formation which relates to the increase in smoke emissions [8]. The lower smoke emissions were found for the combustion of WPO compared to diesel fuel. The reason for reduced smoke for WPO can be early evaporation of fuel leading to premixed or homogeneous charge inside the cylinder [6].

4. Conclusion
Pyrolysis oil or waste plastic crude oil (WPCO) and waste plastic oil through the fractional distillation (WPO) were extracted from waste plastics by pyrolysis process and its chemical composition was characterized by using GC-MS. Later, this study sets out to investigate the effects of WPO in a single cylinder on engine performance of a DI diesel engine was then compared with commercial diesel fuel. The following conclusions were drawn from the investigation.

- The properties of WPCO were found to be closer to gasoline. GC-MS characterization revealed that WPCO consists of 22 chemical compounds with light hydrocarbon chains.
- The properties of WPO were found to be closer to commercial diesel fuel. GC-MS characterization revealed that WPO consists of 17 chemical compounds with heavy hydrocarbon chains.
- The physical and chemical properties of WPCO and WPO were analyzed and compared with commercial diesel fuel and found that it has properties difference to that of diesel fuel for WPCO. While fuel properties of WPO was found to be closer to commercial diesel fuel.
- The use of WPO tended to increase brake specific fuel consumption, resulting in the decrease in brake thermal efficiency when compared with commercial diesel fuel.
- The combustion of waste plastic oil tended to increase NOX while a reduction in smoke was obtained with respect to diesel fuel combustion.
- WPO has been shown a feasible alternative for used as fuel in compression ignition engines.

5. References
[1] Janyalertadun, A., Santaweesuk, C., & Sanongraj, S. (2017). Fuel production, performance, and emission of a CI engine using waste plastics oil. *World Journal of Engineering* **14** 114-120.
[2] Thahir, R., Altway, A., Juliastuti, S. R., & Susianto. (2019). Production of liquid fuel from plastic waste using integrated pyrolysis method with refinery distillation bubble cap plate column. *Energy Reports* **5** 70-77.
[3] Nugroho Pratama, N., & Saptoadi, H. (2014). Characteristics of Waste Plastics Pyrolytic Oil and Its Applications as Alternative Fuel on Four Cylinder Diesel Engines. *International Journal of Renewable Energy Development (IJRED)* **3(1)** 13-20.
[4] Damodharan, D., Sathiyagnanam, A. P., Rana, D., Rajesh Kumar, B., & Saravanan, S. (2017). Extraction and characterization of waste plastic oil (WPO) with the effect of n-butanol addition on the performance and emissions of a DI diesel engine fueled with WPO/diesel blends. *Energy Conversion and Management* **131** 117-126.
[5] Devaraj, J., Robinson, Y., & Ganapathi, P. (2015). Experimental investigation of performance, emission and combustion characteristics of waste plastic pyrolysis oil blended with diethyl ether used as fuel for diesel engine. *Energy* **85** 304-309.
[6] Geo, V. E., Sonthalia, A., Aloui, F., & Femilda Josephin, J. S. (2018). Study of engine performance, emission and combustion characteristics fueled with diesel-like fuel produced from waste engine oil and waste plastics. *Frontiers of Environmental Science & Engineering* **12(4)** 8 .
[7] Sukjit, E., Liplap, P., Maithomklang, S., and Arjharn, W. (2017) "Experimental Investigation on a DI Diesel Engine Using Waste Plastic Oil Blended with Oxygenated Fuels" *SAE Technical Paper* **2017-24-0116**.
[8] Sukjit, E., Herreros, J. M., Piaszyk, J., Dearn, K. D., & Tsolakis, A. (2013). Finding Synergies in Fuels Properties for the Design of Renewable Fuels-Hydroxylated Biodiesel Effects on Butanol-Diesel Blends. *Environmental Science & Technology* **47(7)** 3535-42.

Design of C Band High Power Teflon Water Load

Dingzhen Li[1,2], Lianmin Zhao[1], Fukun Liu[1]

[1] Institute of Plasma Physics, Chinese Academy of Sciences, 230031, Hefei, China
[2] University of Science and Technology of China, 230026, Hefei, China

E-mail: lmzh@ipp.ac.cn

Abstract. A C band Teflon water load is designed for the 4.6 GHz lower hybrid current drive (LHCD) system in the experimental advanced superconducting tokamak (EAST) to absorb the high power microwave reflected from the circulator, Teflon is used to substitute traditional ceramics to isolate the air and the water, which is cheaper and easier to processed. The simulation result carried out by CST microwave studio shows that the return loss reaches -64 dB at 4.6 GHz, and the power loss density was fitted by Origin software to calculate the thermal distribution of the Teflon and the water. Besides, the thermal analysis carried out by ANSYS indicates that this water load can work under 5 kW CW, with the maximum temperature of Teflon less than 71 ℃, and the maximum thermal deformation and stress are no more than 0.14 mm and 3.2 MPa.

1. Introduction

Water load can be used to absorb high power microwave generated by many different high power microwave sources such as klystrons [1-3], magnetrons [4], gyrotrons [5-7], accelerator [8] and some other devices [9-11]. Traditional water loads used to adopt special ceramics, alumina (Al2O3), beryllium oxide (BeO) and aluminum nitride (AlN), to isolate the air and the water or keep the vacuum environment. Compared to these ceramics, Teflon has the similarly excellent microwave performance with low loss tangent (at 25℃). Besides, Teflon is a mature plastic material which is easy to produce and process, its price is only nearly 10 percent of alumina' s price and less than 1 percent of beryllium's or aluminum nitride's price, which shows potential value in microwave engineering. While the coefficient of thermal conductivity and the melting point of Teflon are only 0.25 W/m and 327 ℃, respectively. As a result, it can just work under several kilowatts steady state or at megawatt level for several microseconds.

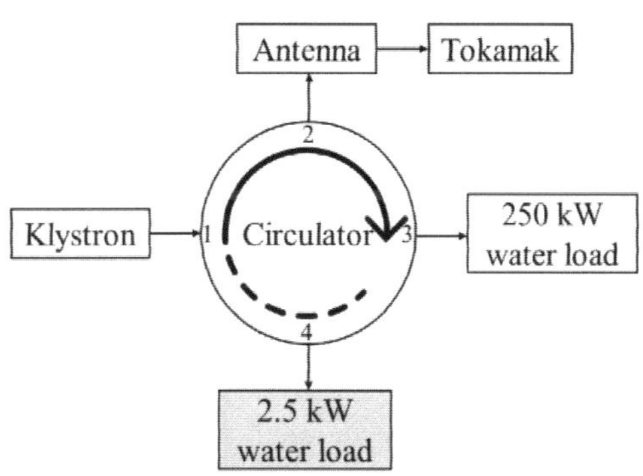

Figure 1. The diagram of 4.6 GHz LHCD system in EAST.

IOP Conf. Series: Materials Science and Engineering **717** (2020) 012002 doi:10.1088/1757-899X/717/1/012002

As shown in Fig.1, in the 4.6 GHz 250kW LHCD system of EAST, each transmission line has 2 water loads connected with the circulator to protect the klystron from the high power reflecting microwave. In the port 3 of the circulator, this water load must have the ability to absorb more than 99 percent reflecting energy, and the rest of the energy (less than 2.5 kW) can be absorbed by water load at port 4. Above all, Teflon water load would be a more affordable structure to be adopted at port 4.
In section II, this work aim to calculate the physical size and the simulate the return loss, power loss density by CST microwave studio. In section III, the power loss density was fitted by Origin to analysis the thermal distribution in the Teflon and the water with * functions. Section IV indicates the thermal analysis results carried out by ANSYS Fluent, which show that the Teflon water load can work under 5 kW CW. Finally, a brief conclusion is given.

2. Calculation of the physical size and microwave simulation

The foremost challenge of the design of the Teflon water load is that Teflon cannot stand high temperature compared with ceramics, and enlarging the propagating cutting plane is a common method to reduce the power loss density. In the 4.6 GHz LHCD system in EAST, all the microwave structures are connected by WR229 rectangular waveguides (58.2 mm × 29.1mm), and the water load cutting plane was expanded to a circle with a radius of 50 mm (see Fig. 2). while the effect is that the microwave would propagate in several different modes.
The cutting wavelengths [12] of circular waveguide can be calculated by

$$\lambda_{\mathrm{cTE}_{mn}} = 2\pi R / p'_{nm} \tag{1}$$

$$\lambda_{\mathrm{cTM}_{mn}} = 2\pi R / p_{nm} \tag{2}$$

where the R is the radius of the circular waveguide and p'_{nm} , p_{nm} refer to the extrema, zeros of the Bessel function of first kind. For 4.6 GHz electromagnetic microwave, it can be calculated that there are 5 different propagation modes: TE_{11}, TM_{01}, TE_{21}, TE_{01} and TE_{31} (arranged in order of the increasing waveguide wavelength λ_g). According to the transmission line theory, the length of the Teflon should be integer multiple of the $\lambda_g/2$, and λ_g can be calculated by

$$\lambda_g = \frac{1}{\sqrt{1/\lambda^2 - 1/\lambda_c^2}} \tag{3}$$

As a result, the minimum waveguide wavelength $\lambda_{g\mathrm{TE}_{11}}$ is 70.6 mm, and the maximum waveguide wavelength $\lambda_{g\mathrm{TE}_{31}}$ is 132 mm. Considering the relative permittivity of the Teflon $\varepsilon_r = 2.08$, so the length of the Teflon should be between 24.48 mm and 45.76 mm. The height and the thickness after optimizing are 26 mm and 41.5 mm, respectively. As shown in Fig. 3, the return loss could reach -64 dB at 4.6 GHz.

3. Fitting power loss density distribution functions

3.1. Power loss density of water

The power loss density of water under the input power of 0.5 W (default) simulated by CST microwave studio is shown in Fig. 4, and the X direction is the propagating direction. It is clear that the power loss density obeys Gauss2D function in YOZ cutting plane, the equal phase plane (shown in x=25mm). And according to the skin effect, it would obey exponential function along the transmission direction. The power loss density of the water was fitted with these functions by Origin software, and the fitted results are reported in Fig. 5. Compared with the central area, power loss densities in other areas are less than one forth, which would be ignored in fitted functions.

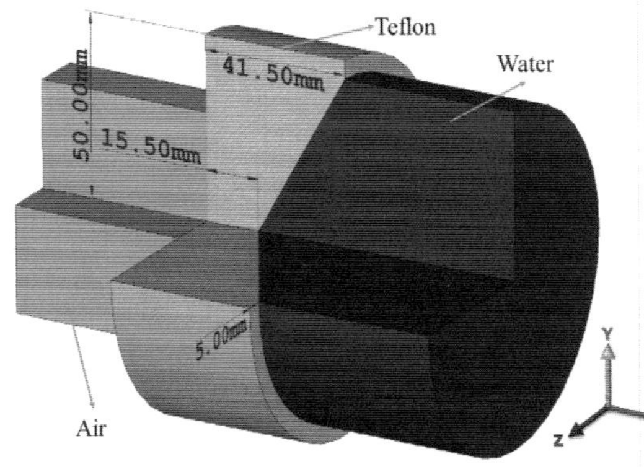

Figure 2. The physical model of Teflon water load (3/4 part).

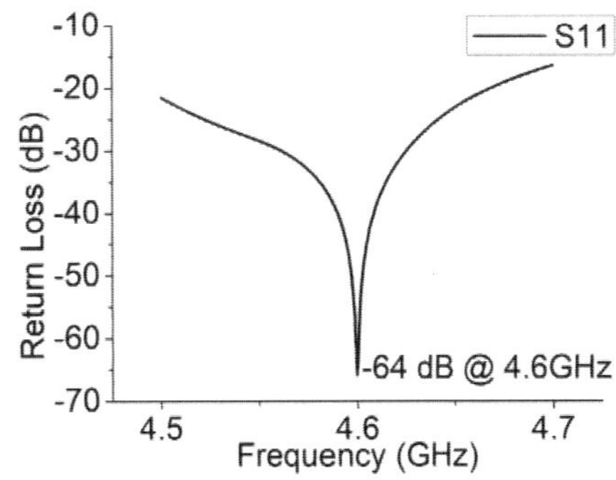

Figure 3. Return loss simulated by CST.

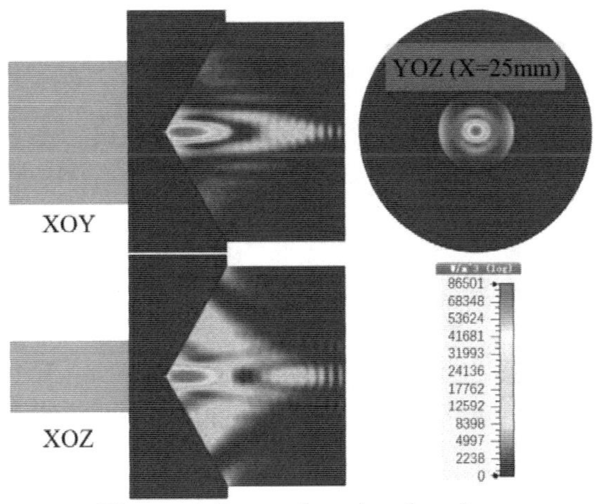

Figure 4. Power loss density of water.

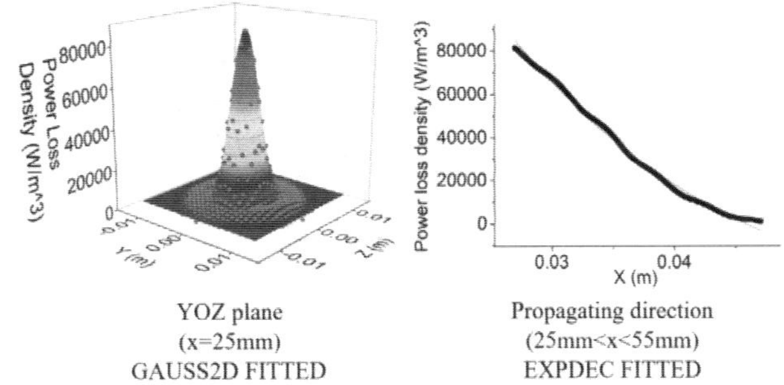

Figure 5. Power loss density functions fitted results.

It is indicated that the Adj. R-Square are 0.99385 and 0.99453 separately, which means that the fitting results are extremely successful. The fitted functions are reported in Table 1.

3.2. Power loss density of the Teflon

Since the power loss density in the Teflon is less than 0.1 percent compared to it in the water, it is hard to observe the distribution functions in the Teflon. So E-field distribution is used to calculated the power loss density distribution in the Teflon. The power loss can be calculated by

$$P = \omega\varepsilon_0\varepsilon_r^{'} \tan\delta|E|^2/2 \tag{4}$$

where the ε_0, $\tan\delta$ are the permittivity of vacuum and loss tangent of the Teflon, and $\omega = 2\pi \times 4.6 \times 10^9$ rad/s,

Table 1. Power Loss Density Fitted Functions

Area	Function type	Expression	Adj. R-Square
YOZ plane (X=25mm)	GAUSS2D	$P_1=z_1+A_1$*exp(-0.5*((Y-x_{c1})/w_1)^2-0.5*((Z-y_{c1})/u_1)^2)	0.99385
Propagating direction (Y=0,Z=0)	EXPDEC	$P_2=A_2$*exp(-X/t_2)+b_2	0.99453
Parameter values	z_1=4154.29856; A_1=83799.42159; x_{c1}=3.23448e-19; w_1=0.00202; y_{c1}=4.60423e-8; w_2=0.00251		
	A_2=541119.31483; t_2=0.01995; b_2=-54617.74196		

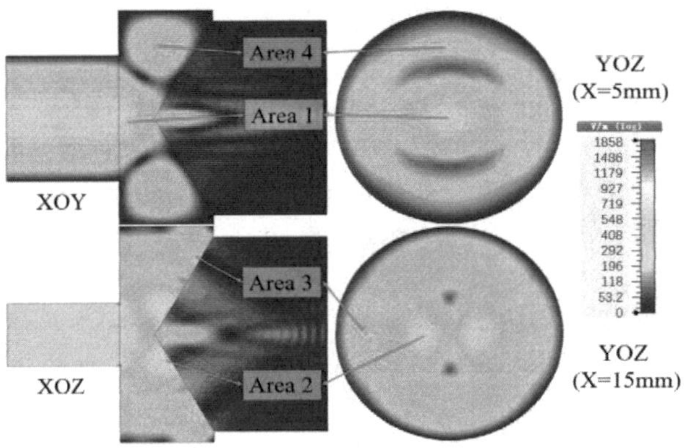

Figure 6. E-field distribution of the Teflon.

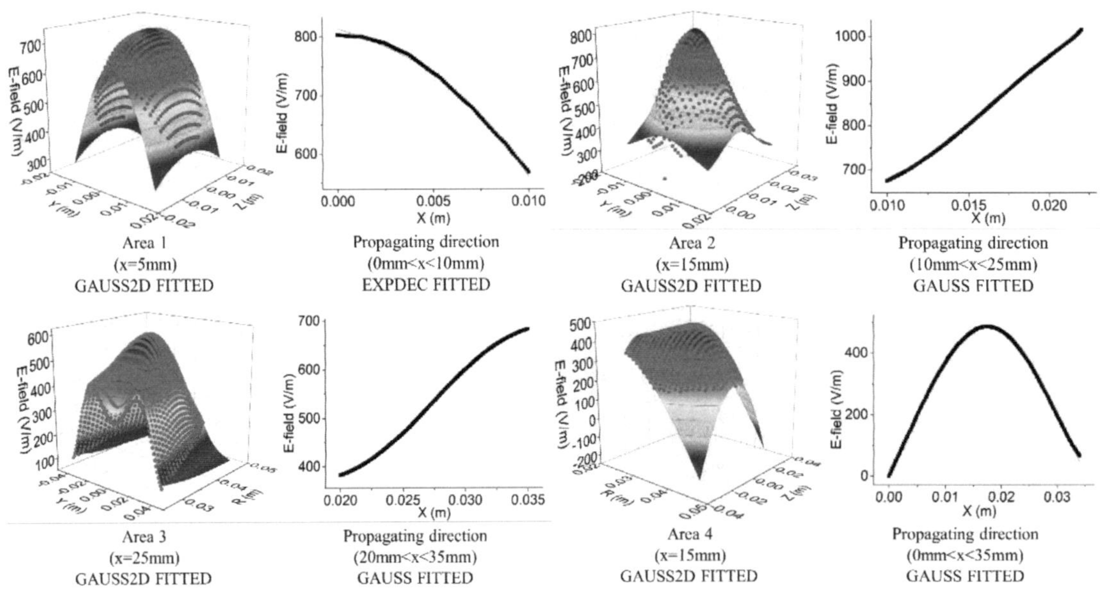

Figure 7. E-field distribution functions fitted results.

Table 2. E-field Distribution Fitted Functions

Area	Function type	Expression	Adj. R-Square
YOZ plane Area 1 (X=5mm)	GAUSS 2D	$E_3=z_3+A_3*exp(-0.5*((Y-x_{c3})/w_3)^2-0.5*((Z-y_{c3})/u_3)^2)$	0.99699
Propagating direction 1 (Y=0,Z=0)	EXPDEC	$E_4=A_4*exp(-X/t_4)+b_4$	0.99652
YOZ plane Area 2 (X=15mm)	GAUSS 2D	$E_5=z_5+A_5*exp(-0.5*((Y-x_{c5})/w_5)^2-0.5*((Z-y_{c5})/u_5)^2)$	0.9701
Propagating direction 2 (Y=0,Z=0)	GAUSS	$E_6=b_6+(A_6/(w_6*sqrt(pi/2)))*exp(-2*((X-x_{c6})/w_6)^2)$	0.99994
YOZ plane Area 3 (X=25mm)	GAUSS *GAUSS	$E_7=(b_7+(A_7/(w_7*sqrt(pi/2)))*exp(-2*((sqrt(Y^2+Z^2)-x_{c7})/w_7)^2)*(c_7+(k_7/(u_7*sqrt(pi/2)))*exp(-2*((Y-y_{c7})/u_7)^2))/600$	0.99169
Propagating direction 3 (Y=0,Z=37mm)	GAUSS	$E_8=b_8+(A_8/(w_8*sqrt(pi/2)))*exp(-2*((X-x_{c8})/w_8)^2)$	0.99989
YOZ plane Area 4 (X=15mm)	GAUSS *GAUSS	$E_9=(b_9+(A_9/(w_9*sqrt(pi/2)))*exp(-2*((sqrt(Y^2+Z^2)-x_{c9})/w_9)^2)*(c_9+(k_9/(u_9*sqrt(pi/2)))*exp(-2*((Z-y_{c9})/u_9)^2))/470$	0.99831
Propagating direction 4 (Y=33mm,Z=0)	GAUSS	$E_{10}=b_{10}+(A_{10}/(w_{10}*sqrt(pi/2)))*exp(-2*((X-x_{c10})/w_{10})^2)$	0.99938
Parameter values		z_3=-615.14344; A_3=1360.83596; x_{c3}=4.6789e-19; w_3=0.01824; yc_3=0.58908e-7; u_3=0.03783	
		A_4=-58.85349; t_4=-0.00598; b_4=872.69072	
		z_5=295.13611; A_5=527.65302; x_{c5}=-3.28815e-19; w_5=0.00907; y_{c5}=0.01279; u_5=0.00731	
		b_6=1182.21655; A_6=-14.13536; w_6=0.02083; x_{c6}=0.00622	
		b_7=137.92733; A_7=7.20801; w_7=0.01457; x_{c7}=0.03693; c_7=233.24537; k_7=24.77161; u_7=0.0448; y_{c7}=-7.75368e-10	
		b_8=331.0194; A_8=6.99422; w_8=0.01588; x_{c8}=0.03579	
		b_9=-627.41772; A_9=45.64531; w_9=0.04405; x_{c9}=0.0332; c_9=912.70274; k_9=8.19763; u_9=0.03381; y_{c9}=3.13092e-6	
		b_{10}=-402.68034; A_{10}=31.47404; w_{10}=0.02828; x_{c10}=0.0175	

4. Fluid and thermal analysis

In the actual engineering design, it is necessary to preserve the margin to ensure the safety situation. For 4.6 GHz LHCD system in EAST, the power capacity of the water load must reach 5 kW at the port 4 of the circulator. Taking the requirements of the 4.6 GHz LHCD water system in EAST into consideration, the radius of the water pipe is 10 mm, and the velocity of the water at the inlet should be less than 2 m/s to reduce the mechanical noise. According to formula

$$\Delta T = \frac{P}{c \cdot S \cdot v \cdot \rho} \tag{5}$$

the temperature increase ΔT would be only 2 ℃ under the 5 kW input power, where P is the input power, c is the specific heat capacity of water, S is the cross-sectional area of pipe, v is the velocity of the water, and ρ is the density of water. However, the thermal conductivity of Teflon is only 0.25 W/m, which would cause there is a huge temperature difference between the Teflon and the water, since that the heat generated in the Teflon can hardly be transferred to the water.

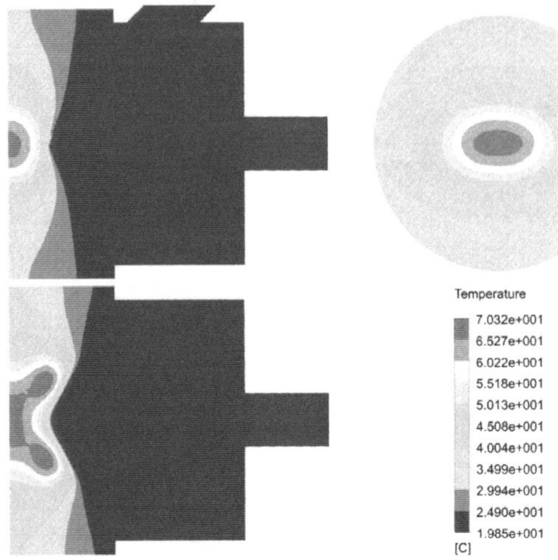

Figure 8. Heat distribution of Teflon water load under input power of 5 kW.

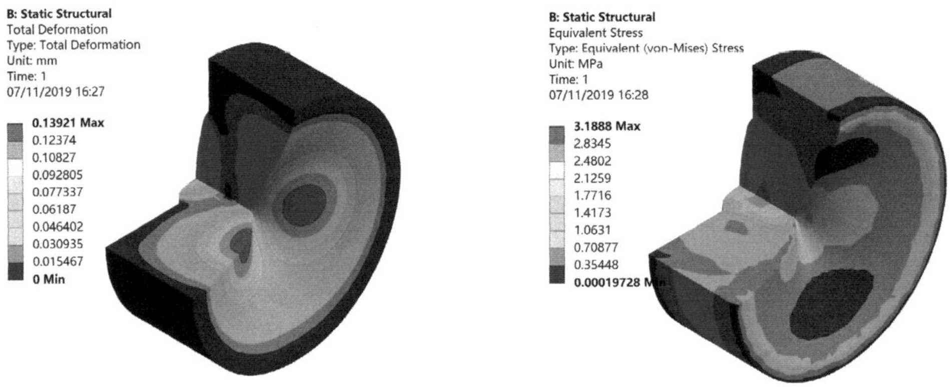

(a) (b)

Figure 9. Thermal analysis results carried out by ANSYS Static Structural.
(a) Deformation; (b) Stress.

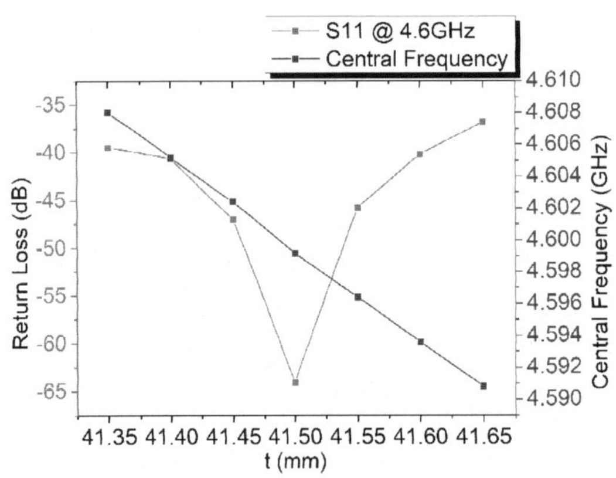

Figure 10. Return loss @ 4.6 GHz and the central frequency changing with the parameter "*t*"

Adjusting the angle of the input water pipe to make the water take the heat generated in the water away rapidly, setting the inlet velocity, outlet pressure and the room temperature to 2 m/s, 0.5MPa and 20 °C separately, multiplying the fitted functions by 10000 to get UDF (user defined function) under input power of 5 kW and compiling the UDF in ANSYS Fluent to simulate the heat distribution of the Teflon water load. Simulated result is given in Fig. 8. It is indicated that the maximum temperature of the water load is 70.32 °C, located at the center of the Teflon. Importing these results to ANSYS Static Structural Unit to get the thermal deformation and stress, it can be observed that maximum deformation and stress are only 0.139 mm and 3.19 MPa, respectively, see Fig. 9. The tensile yield strength of Teflon is 20 MPa [13], which means that the Teflon has the ability to stand the stress under input power of 5 kW. The parameter "*t*" (total thickness of the Teflon) sweeping result is given in Fig. 10, which shows that the microwave performance is almost unaffected by the error of 0.15 mm (maximum thermal deformation).

5. Conclusion

A 5 kW Teflon water load is designed for the 4.6 GHz LHCD system in EAST, taking the microwave theory and fluid thermal analysis into consideration simultaneously. The UDF was written by fitting the distribution of the E-filed and the power loss density, final simulated results indicate that this water load can absorb more than 99 percent energy under 4.6 GHz 5 kW microwave with little thermal deformation and sustainable stress.

6. References

[1] Liu L, Liu F K, Shan J F and Kuang G L 2007 Design of a New Water Load for S-band 750 kW Continuous Wave High Power Klystron Used in EAST Tokamak Plasma Science and Technology 9 223-6
[2] Koert P, MacGibbon P, Beck W, Doody J and Gwinn D 2007 High Power Water Load for Lower Hybrid Current Drive at 4.6 GHz on Alcator C-Mod. In: Symposium on Fusion Engineering, pp 1-4
[3] Horan D, Kustom R and Ferguson M 1995 Performance of a 2-megawatt high voltage test load. In: Proceedings Particle Accelerator Conference, pp 1988-90
[4] Gao D P, Zhang R, Ding H B and Tang K 2015 Design of 100kW Water Load for P-band Continuous Wave Magnetron. In: International Vacuum Electronics Conference, (Beijing: IEEE) pp 27- 9
[5] Illy S, Gantenbein G, Schmid M and Weggen J 2011 Design, construction and first tests of a stainless steel load for high power MM-wave radiation. In: 2011 Abstracts IEEE International Conference on Plasma Science, pp 1
[6] Ives R L, Mizuhara M, Schumacher R and Pendleton R 1999 Development of a high power CW waterload for Gaussian mode gyrotrons IEEE Transactions on Plasma Science 27 531-7
[7] Ok J-W, Choi S, Lee B-S, Yoon J-H, Park J Y, Shin C S, Won M-S and Hong I-S 2015 Development of a water load for a Ka-band 10 kW continuous-wave gyrotron in an ECR ion source system Journal of the Korean Physical Society 66 349-52
[8] Gold S H and Gai W 2001 High-power testing of 11.424-GHz dielectric-loaded accelerating structures. In: Proceedings of the 2001 Particle Accelerator Conference pp 3984-6 vol.5
[9] Panahi M A, AbdollahRamezani S, Ebrahimzadeh Ardakani G and Safian R 2015 An Efficient High Power RF Dummy-Load IEEE Microwave and Wireless Components Letters 25 409-11
[10] Han J, Kim J-h, Park S-d, Yoon M, Park S Y, Choi D W, Shin J W and So J H 2009 Coaxial-type water load for measuring high voltage, high current and short pulse of a compact Marx system for a high power microwave source Physical Review Special Topics - Accelerators and Beams 12
[11] Matsumoto H, Iino Y, Fujiwara C, Kabeya Z and Onda T 1999 Experience on the high-power SiC microwave dummy-load using SiC absorber. In: Proceedings of the 1999 Particle Accelerator Conference, pp 842-4 vol.2
[12] Pozar D M 2005 Microwave Engineering vol 3 (USA: Wiley)
[13] Chin G Y, Nesbitt E A and Williams A J 1966 Anisotropy of strength in single crystals under plane strain compression Acta Metallurgica 14 467-76

Acknowledgments
This work was supported by National MCF Energy R&D Program of China under Grant 2018YFE0305100 and National Natural Science Foundation of China under Grant 11675212.

IOP Conf. Series: Materials Science and Engineering **717** (2020) 012003 doi:10.1088/1757-899X/717/1/012003

Crude Castor Oil as Blend Component in Diesel/Ethanol Fuel Blend: Combustion Characteristics and Exhaust Emissions

Niti Klinkaew, Ekarong Sukjit and Teetut Dolwichai

School of Mechanical Engineering, Institute of Engineering, Suranaree University of Technology, Nakhon Ratchasima 30000, Thailand

E-mail: ekarong@sut.ac.th

Abstract. Some properties such as poor lubricity and low viscosity limit the use of ethanol as an alternative fuel in diesel engines. To improve such inferior properties, crude castor oil with excellent lubricity and extremely high viscosity was added to diesel-ethanol fuel blend. Investigation on combustion characteristics, engine performance and exhaust emissions of a four-stroke single cylinder diesel engine fueled with ternary blend of diesel-ethanol-castor oil was carried out in this research. The blend of 80% diesel, 10% ethanol and 10% castor oil (D80E10C10) was selected for the study. The results on fuel properties showed that the presence of 10% castor oil in diesel-ethanol blend can improve the properties of the blend and meet the diesel fuel specification. The delay in start of combustion was obtained with the use of the ternary blend with respect to diesel fuel combustion, resulting in the lower peak of in-cylinder pressure and temperature. The reduction in nitrogen oxides (NOx) and smoke was found when the engine was operated with the ternary blend at all conditions tested. Carbonaceous gas emissions, carbon monoxide (CO) and unburnt hydrocarbons (HC), was increased with the combustion of ternary blend at low engine operating loads. However, the difference in CO and HC with the combustion of diesel fuel and ternary blend was decreased as the engine loads increased. The addition of crude castor oil to diesel-ethanol blend used as alternative fuel in the diesel engine did not show the significant difference in brake specific fuel consumption and brake thermal efficiency compared to diesel fuel.

1. Introduction

Nowadays, the resources of petroleum used as fuel are depleting with increasing demand, as well as increasingly stringent regulations, pose a challenge to Thailand, which is the non-oil producing country. With the commercialization of bioenergy, it has provided an effective way to fight against the problem of petroleum scarcity and the influence on environment. Although there is an increase in number of literatures to study engine performances and exhaust emissions when using biodiesel, especially in this decade, advanced biofuels such as bio-alcohol and second-generation biodiesel which are derived from waste, agriculture residues and non-food crops have been receiving more attention as sustainable alternative to fossil fuels. Oxygen present in fuel molecules of bio-alcohol and biodiesel can participate in a cleaner combustion process, resulting in a reduction of engine-out emissions to meet increasingly stringent vehicle emission standards. Previous publications have studied the use of diesel fuel blended with ethanol in diesel engines and the results found that the addition of lower 20% ethanol to diesel fuel can be used in diesel engines without any engine modification [1]. It was found that the presence of ethanol in fuel blend participates in the reduction in carbon monoxide [2-3]. More emission benefits of ethanol was obtained due to higher heat of vaporization which tended to reduce combustion temperature, being less favorable NOx formation [4]. However, the use of ethanol with high percentage as blend component in diesel fuel is not compatible with fuel system and engine components. Fuel containing with high proportion of ethanol is need to improve some key properties. To increase amount of advanced fuel used in vehicles and restore some inferior properties of diesel-ethanol fuel blend, biodiesel derived from castor oil with excellent lubricity and extremely high viscosity was studied [5]. The addition of 10%-15% castor oil methyl ester was enough to improve the lubricity of the diesel-ethanol fuel blend while its viscosity was under the limit of diesel fuel specification. The presence of castor oil methyl ester also increased calorific value, cetane number and flash point of diesel-ethanol blend. Moreover, hydroxyl group belonging to ricinoleic acid (C18:1OH) which is main fatty acid present in castor oil can enhance soot oxidation process.

To extend the benefit of castor oil and be more convenient for applying as blend component, crude castor oil was added to diesel-ethanol fuel blend in this work. The physical and chemical properties of ternary blend (diesel-ethanol-castor oil) were studied along with combustion and engine-out emissions and the engine operating with diesel fuel was used as a baseline test.

2. Materials and methods

2.1. Test fuels

The presence of 10% of anhydrous ethanol by volume in the diesel fuel was selected to study the effect of castor oil on fuel properties, combustion characteristics and engine-out emissions. With preliminary work on lubricity experiment, the addition of 10% castor oil was enough to restore the lubricity of the ethanol blends, there was no significant improvement on the blend lubricity after this percentage of castor oil. Consequently, the fuel blend of 10% castor oil, 10% ethanol and 80% diesel fuel (D80E10C10) was used to operate with the engine. The basic physical and chemical properties of test fuels was measured according to ASTM standards and was shown in Table 1. The fatty acid profile of castor oil was detailed in Table 2 and it can be confirmed that the majority of fatty acid present in castor oil is ricinoleic acid with more than 85% wt.

Table 1. Physical and chemical properties of test fuels.

Fuel Properties	Units	Test Method	Diesel	Ethanol	Castor oil	D80E10C10
Kinematic viscosity at 40 °C	cSt	ASTM D445	3.54	1.26	289	4.06
Flash point	°C	ASTM D93	78	13.5	282	158
Specific gravity at 15.6 °C	-	ASTM D1298	0.828	0.785	0.950	0.836
Density at 15.6 °C	kg/m^3	ASTM D1298	827.2	784.2	949.1	835.2
Cetane index	-	ASTM D976	60.18	8^{CN}	48^{CN}	55.2
Gross Calorific Value	(MJ/kg)	ASTM D240	45.39	26.83	36.16	42.23

Table 2. Fatty acid profile of castor oil. (%wt).

Fatty acid		% wt
Lauric	C12:0	0.02
Myristic	C14:0	0.06
Palmitic	C16:0	1.63
Stearic	C18:0	1.66
Oleic	C18:1	3.85
Ricinoleic	C18:1OH	85.6
Linoleic	C18:2	6.04
Linolenic	C18:3	0.43
Arachidic	C20:0	0.08
Gadoleic	C20:1	0.58
Eicosadienoic	C20:2	0.05

2.2. Experimental setup

The engine used for this study is a four-stroke, single-cylinder, water-cooled, direct injection compression ignition engine. An eddy current dynamometer with a load cell was used to load the engine. The tests were performed at rated speed of 1500 rpm and three different engine loads (25%, 50% and 75% of maximum engine torque). Commercial diesel fuel was used as a baseline comparison with waste plastic oil for the engine run tests. An air box was used to measure the air flow rate to the engine and volumetric fuel flow rate was measured using a burette and stopwatch. A pressure transducer (mounted on the cylinder head) is connected to a charge amplifier which in turn is connected to computer and a crank angle encoder is used to find crank angle with 1o revolution. The schematic diagram of the test engine setup is given in Figure 1 and the technical specifications of the engine are given in Table 3. The TESTO 350 analyzer was employed to measure nitrogen oxide (NOx), carbon monoxide (CO), hydrocarbon (HC) and TESTO 308 was used to evaluate smoke index. The specification of the gas analyzing device shown in Table 4. At every load, readings were taken after the engine reached steady state. For every refueling, engine was kept running for 10 minutes such that it consumes entire fuel entangled in fuel pipe lines. Repeatability of readings was guaranteed by duplicating the investigations thrice.

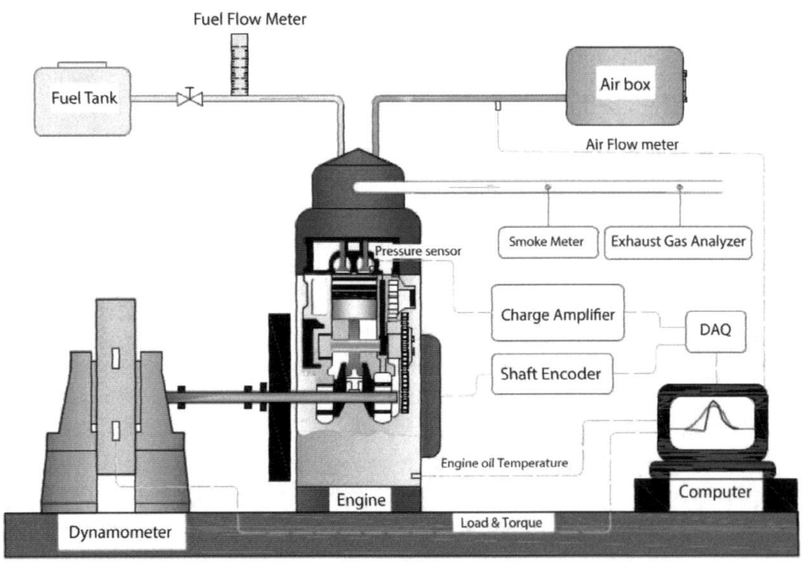

Figure 1. I Schematic diagram of the engine test setup.

Table 3. Technical specifications of the engine.

Engine	Specification
Model	Kirloskar TV1
Engine type	1 cylinder, 4 strokes, Water cooled, Direct injection
Bore	87.5 mm
Stroke	110 mm
Swept volume	661 cc
Rate output	3.5 kW at 1500 rpm
Compression ratio	12:1-18:1
Start of injection	23 ^{o}CA bTDC
Injection pressure	210 bars
Dynamometer	Eddy current

Table 4. Specification of gas analysing devices.

Measured quantity	Measuring range	Resolution	Accuracy
TESTO 350			
NO	0 to 4000 ppm	1 ppm	± 5 ppm
NO_2	0 to 500 ppm	0.1 ppm	± 5 ppm
CO	0 to 10000 ppm	1 ppm	± 5 ppm
HC	100 to 21000 ppm	10 ppm	± 400 ppm
TESTO 308			
Smoke index	0 to 6	0.1	± 0.2

3. Results and discussion

3.1. Engine performance

In the present investigation, experiments were carried out to understand engine performance, combustion characteristics and exhaust emissions of the diesel engine by using diesel fuel and the ternary fuel blend (D80E10C10). The blend was used and the experimental observations were collected so that the results were compared with that of diesel fuel. The engine was loaded in the range of 25%, 50%, and 75% load for a constant speed of 1500 rpm. Figure 2 shows the relation between applied loads and the brake specific fuel consumption. It can be seen that the brake specific fuel consumption decreased as the engine operating loads increased. This was due to the fuel was converted into thermal energy that improved efficiency. The lower brake specific fuel consumption implies that less amount of fuel is needed to generate unit power output. Although, some pore properties of the ternary blend such as higher viscosity and lower calorific value tend to increase brake specific fuel consumption. The better lubricating properties of castor oil can play a role to reduce the power lost by friction [6]. Inversely, brake thermal efficiency increased as the engine operating loads increased (Figure 3). Brake thermal efficiency indicates the ability of combustion system to accept the experimental fuel and provides a comparable means of assessing how efficiently the fuel was converted into mechanical output [7-8]. Brake thermal efficiency increased as engine load increased because more fuel was dispensed with increasing combustion temperature, resulting in high thermal efficiency. Comparing diesel fuel and the ternary blend, the brake specific fuel consumption and brake thermal efficiency was no significant difference between both fuels tested.

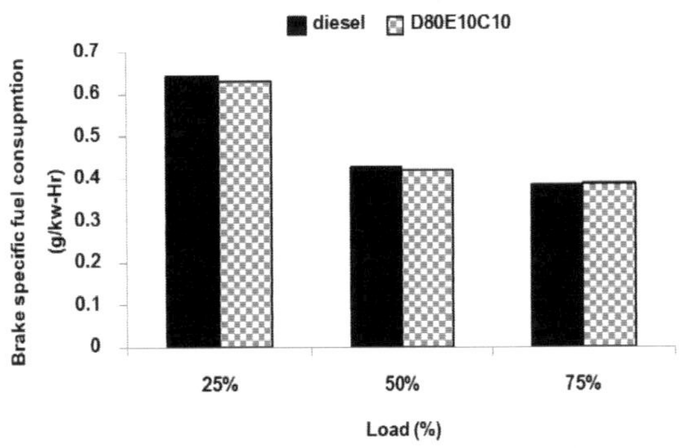

Figure 2. Brake specific fuel consumption.

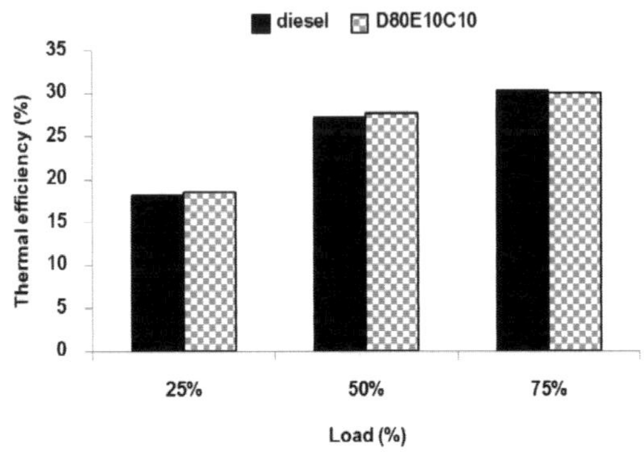

Figure 3. Brake thermal efficiency.

3.2. Combustion characteristics

In-cylinder pressure (ICP) and rate of heat release (ROHR) for all engine operating loads are depicted in Figure 4. The first law of thermodynamics was used to calculate the heat release rate with equations (1).

$$\frac{dQ}{d\theta} = \frac{\gamma}{\gamma\text{-}1} p \frac{dV}{d\theta} + \frac{1}{\gamma\text{-}1} V \frac{dp}{d\theta} \tag{1}$$

where $dQ/d\theta$ is the rate of heat release, p is the in-cylinder pressure, V is the combustion chamber volume, and θ is the crank angle. The constant specific heat ratio (γ) of 1.35 was used for the calculation. 100 consecutive engine cycles were performed in order to analyse the in-cylinder pressure and rate of heat release.

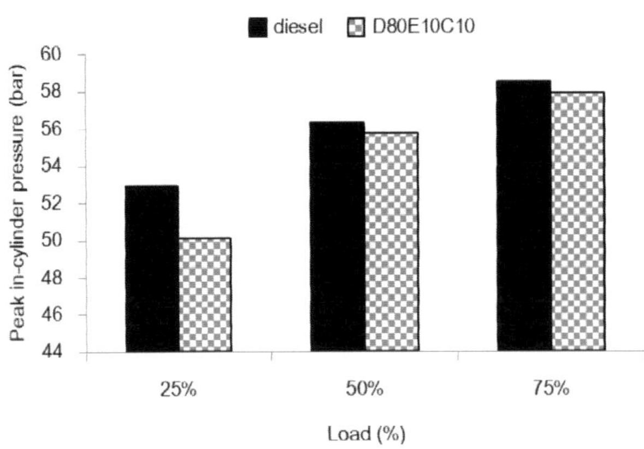

Figure 4. Peak of in-cylinder pressure.

Figure 5 shows the combustion characteristics of test fuels. Longer ignition delay (retard in start of combustion) was obtained when the engine was operated with the ternary blend. Low cetane number of ethanol and castor oil together with high heat of vaporization of ethanol can be used to justify the increase in the ignition delay by the combustion of the ternary blend. In addition, lower calorific value of both ethanol and castor oil can participate in lower energy released by the combustion, resulting in lower combustion temperature and in-cylinder pressure. Lower pressure rise in combustion chamber resulted in lower peak of heat release rate.

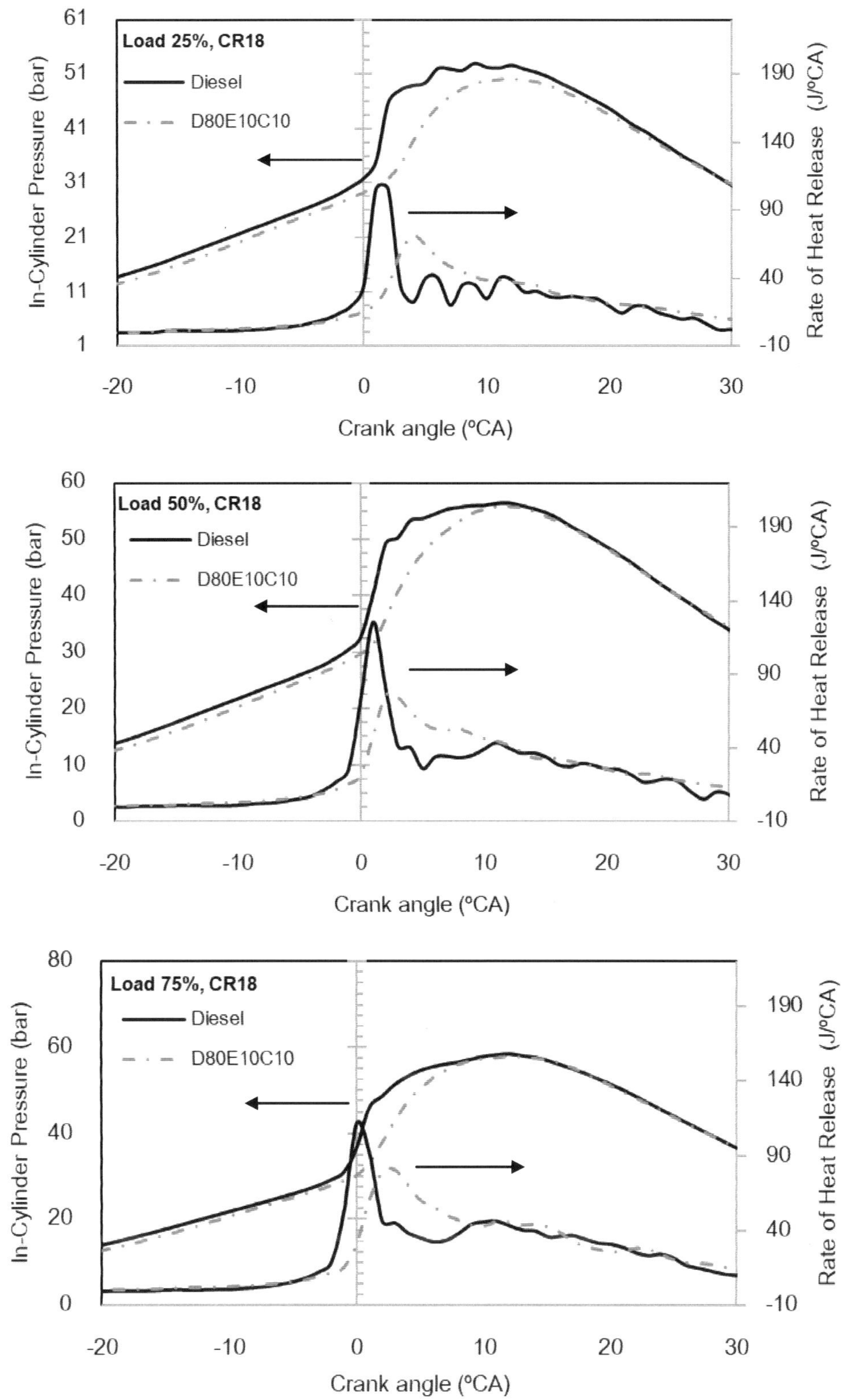

Figure 5. In-cylinder pressure and rate of heat release.

3.3. Emissions

The formation of nitrogen oxides depends on the temperature in the combustion chamber of the engine. The higher combustion temperature, the higher oxides of nitrogen will be produced. [9] There are also some factors that affect the formation of nitrogen oxides. Combustion of oxygenated fuels such as biodiesel with oxygen contained in fuel molecule, resulting in higher combustion temperature. Low cetane value of fuel causes a long ignition delay, affecting the maximum heat release rate during the combustion of pre-mixed mixtures (premixed combustion). The reduction of oxides of nitrogen can occur by burning fuels with high latent heat of vaporization, such as ethanol, because the fuel that is injected in the combustion chamber will absorb heat to converting liquid to gas phase, resulting in a decrease in combustion temperature In addition, the high viscosity also affects the temperature in the combustion chamber because the fuel injection into the combustion chamber is a large droplet which leads to lower the temperature in the combustion chamber, leading to lower nitrogen oxides [10]. The results in Figure 5 shows that the combustion of ternary blend produced lower oxides of nitrogen at all engine operating loads. This is mainly due to the higher heat of vaporization of ethanol and high viscosity of castor oil. It is notable that the effect of oxygen content which tends to increase the combustion temperature and be more favorable the formation of nitrogen oxides, was compensated with the effect of low heat of vaporization and high viscosity. Carbon monoxide emissions are dependent on properties of fuel, amount of oxygen, combination of air and fuel, temperature and turbulent air flow in combustion chamber. The results in Figure 7 shows that the slight increase in carbon monoxide was obtained with the use of ternary blend at low and mid engine loads while the reduction in carbon monoxide was found at high engine load, compared to diesel fuel. At low load, temperature in combustion chamber may not be high enough to compensate the effect of high heat of vaporization and high viscosity of fuel blend, resulting in low combustion temperature and less complete combustion. When the engine was operated at high load and temperature in combustion chamber was enough to compensate such effect, lower C/H ratio of ethanol may dominate and lead to the reduction in carbon monoxide by the combustion of ternary blend. Figure 8 shows the variation of unburnt hydrocarbon emissions with engine loads. Hydrocarbons emissions are organic compounds that are caused by incomplete combustion. Higher hydrocarbon was found as engine load increased. This is due to more fuel injected to produce more engine torque and power, resulting in higher possibility of fuel not being burned. Comparing the two test fuels, it was observed that the combustion of ternary blend produced higher hydrocarbon emissions than that of diesel fuel at all engine operating loads. This is likely to be a consequence of high heat vaporization and high viscosity of fuel blend which tends to increase incomplete combustion. Smoke measurement is an indirect way of measuring of diesel particulate emissions. The results in Figure 9 shows that the reduction in smoke index was obtained by the combustion of ternary blend compared to diesel fuel at all engine operating loads. The main reason which can be used to support such reduction is the presence of oxygen in fuel molecules of ethanol and castor oil. Hydroxyl group belonging to both ethanol and ricinoleic acid of castor oil was reported to enhance the soot oxidation and inhibit soot formation [11-12].

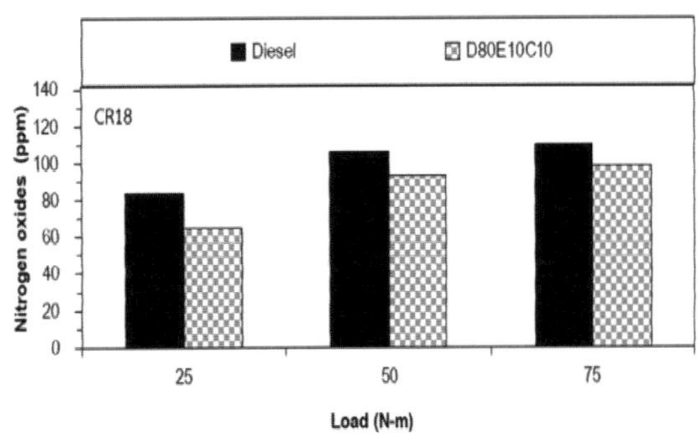

Figure 6. Oxides of nitrogen

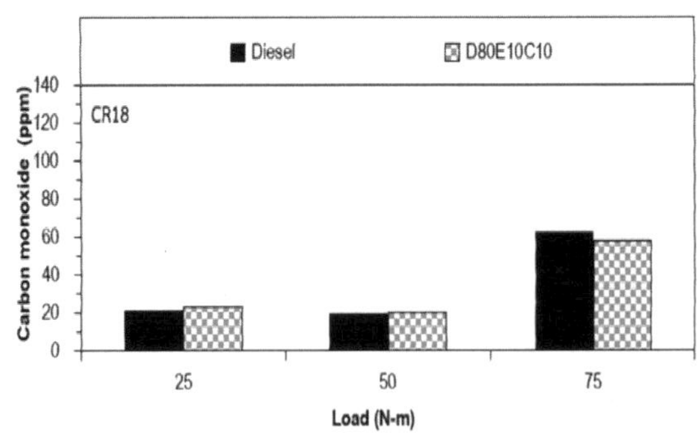

Figure 7. Carbon monoxide

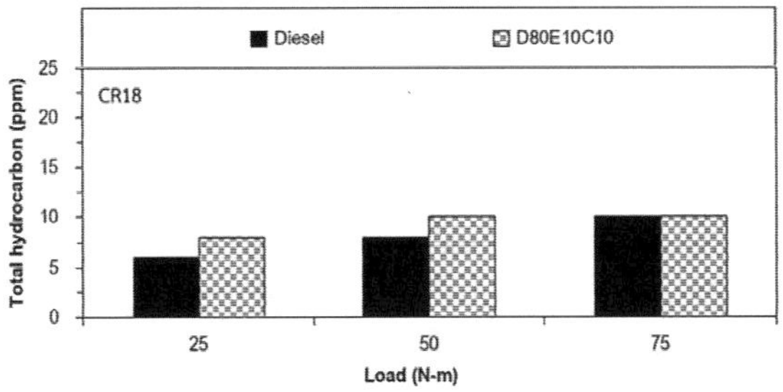

Figure 8. Unburnt hydrocarbons

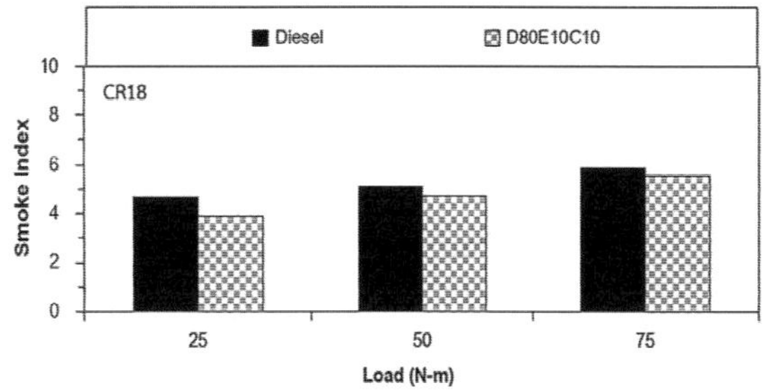

Figure 9. Smoke

4. Conclusions

The study of the use of castor oil with ethanol as an alternative to reducing the use of diesel fuel can be summarized as follows.

- Excellent lubricity and high viscosity of crude castor oil can compensate the loss of such properties of ethanol and the synergic effect was obtained in the study.
- Slight improvement in brake specific fuel consumption and brake thermal efficiency was found when the engine was operated with the ternary blend.
- The combustion of ternary blend showed the delay in start of combustion, lower pressure rise in combustion chamber and lower peak of heat release rate, compared to diesel fuel.
- The benefit in nitrogen oxides and smoke emissions was obtained when the engine was operated with ternary blend at all conditions tested.
- The increase in unburnt hydrocarbon and carbon monoxide emissions were undesirable results for using the ternary blend, especially at low engine operating load.
- Footnotes should be avoided whenever possible. If required they should be used only for brief notes that do not fit conveniently into the text.

5. References

[1] Ajav EA, Bachchan S and Bhattacharya TK 1999 Experimental study of some performance parameters of a constant speed stationary diesel engine using ethanol-diesel blends as fuel *Biomass and Bioenergy* 17(4) pp 357-365

[2] Lapuerta M, Herreros JM, Lyons LL, García-Contreras R and Briceño Y 2008 Effect of the alcohol type used in the production of waste cooking oil biodiesel on diesel performance and emissions *Fuel* 87(15-16) pp 3161-3169

[3] Rakopoulos DC, Rakopoulos CD, Giakoumis EG, Dimaratos AM and Kyritsis DC 2010 Effects of butanol-diesel fuel blends on the performance and emissions of a high-speed DI diesel engine *Energy Conversion and Management* 51(10) pp 1989-1997

[4] Rakopoulos DC, Rakopoulos CD, Papagiannakis RG and Kyritsis DC 2011 Combustion heat release analysis of ethanol or n-butanol diesel fuel blends in heavy-duty DI diesel engine *Fuel* 90(5) pp 1855-1867

[5] Sukjit E, Herreros JM, Dearn K and Tsolakis A 2014 Improving ethanol-diesel blend through the use of hydroxylated biodiesel *SAE Technical Paper* 2014-01-2776

[6] Das M, Sarkar M, Datta A and Santra AK 2018 An experimental study on the combustion, performance and emission characteristics of diesel engine fuelled with diesel-castor oil biodiesel blends *Renewable Energy* 119 pp 174-184

[7] Devaraj J, Robinson Y and Ganapathi P 2015 Experimental investigation of performance, emission and combustion characteristics of waste plastic pyrolysis oil blended with diethyl ether used as fuel for diesel engine *Energy* 85 pp 304-309

[8] Sayin C, Ilhan M, Canakci M and Gumus M 2009 Effect of injection timing on the exhaust emission of a diesel engine using diesel methanol blends *Renewable Energy* 34 pp 1261-1269

[9] Panwar NL, Shrirame HY, Rathore NS, Jindal S and Kurchania AK 2010 Performance evaluation of a diesel engine fueled with methyl ester of castor seed oil *Applied Thermal Engineering* 30(2) pp 245-249
[10] Xuea J, Grift TE and Hansena AC 2011 Effect of biodiesel on engine performances and emissions *Renewable and Sustainable Energy Reviews* 15(2) pp 1098–1116
[11] Sukjit E, Herreros JM, Dearn KD, García-Contreras R and Tsolakis A 2012 The effect of addition of individual methyl esters on the combustion and emissions of etanol and butanol-diesel blends *Energy* 42 pp 364-374
[12] Sukjit E, Herreros JM, Piaszyk J, Dearn KD and Tsolakis A 2013 Finding synergies in fuels properties for the design of renewable fuels – hydroxylated biodiesel effects on butanol-diesel blends *Environmental Science & Technology* 47 pp 3535-3542

Acknowledgments
Suranaree University of Technology (SUT), Nakhon Ratchasima, Thailand and National Research Council of Thailand are gratefully acknowledged for the financial support to this research.

IOP Conf. Series: Materials Science and Engineering **717** (2020) 012004 doi:10.1088/1757-899X/717/1/012004

Design and Optimization of a Wearable upper Limb Exoskeleton based on Adams

Peng Yin[1], Liang Yang[1], Bin Liu[2]

[1] School of Mechanical and Automotive Engineering, South China University of Technology, Guangzhou 510640, China
[2] Development Department, Guangzhou Hyetone Electromechanical Equipment Co., Ltd., Guangzhou 510407, China

Corresponding author's e-mail address: 18814113085@163.com (Peng Yin), leonyounggod@msn.com (Liang Yang), liubin@hyetone.com (Bin Liu).

Abstract. A wearable upper limb exoskeleton for factory assembly positions was designed. The trajectory and dynamics of the worker's arm during assembly was analysed. According to the power demand, the corresponding torque drive was designed at the shoulder joint. Then the corresponding scheme was designed. The effect on joint drive force was simulated on Adams. As shown by the result, pre-compression force and spring stiffness have a great influence on the driving effect. Finally, with the work of additional required force moment for the shoulder joint as the objective function, the two parameters of spring stiffness and pre-compression force were optimized. Compared to not wearing the exoskeleton, the energy consumption of the exoskeleton was reduced by 90%.

1. Introduction

As a way to enhance human strength and assist human body training, Exoskeleton is widely used in military, medical, rescue and daily life [1-4]. The types of exoskeleton mainly include military exoskeletons [5], lower limb exoskeleton [6], Waist-assistive exoskeleton [7], upper limb exoskeleton [8], etc. However, most of the exoskeleton is a human-machine intelligent system integrating mechanical, electrical, sensor and space control [9, 10]. It is expensive, complicated in design and large weight and it is difficult to meet the demand for the current market. Especially for modern manufacturing industrial assembly, workers need to wear and move quickly for a long time, which is obviously not suitable for the job. Based on the analysis of kinematics and dynamics of upper limb joints during assembly work, a lightweight and flexible wearable upper limb exoskeleton is designed.

2. Trajectory of upper limb

The upper limb exoskeleton should always follow the spatial position of the human arm and provide matching assistance as much as possible to meet the requirements of the worker. Therefore, it is important to calculate the trajectory and power demand of the upper limb bones during the assembly work. As shown in Figure 1, from top to bottom are the three positions of the handing tool operation, swing arm after operation and swing arm completion. When handing tool operation, the forearm is at an angle of 90 degrees to the upper arm and the upper arm is at an angle of 90 degrees to the vertical. When swing arm completion, the forearm is at an angle of 180 degrees to the upper arm and the upper arm is at an angle of 30 degrees to the vertical. During the swing arm, the angle between the forearm and the upper arm and the angle between the upper arm and the vertical direction change uniform.

3. Dynamics of upper limb

During the swing arm, the shoulder and elbow joints are mainly used to withstand the force moment generated by the gravity of the tool and arm. The wrist joint withstands less torque due to the small force of the palm. According to the measurement, a male with a height of 170 cm has an upper arm length of about 240 mm, an upper arm mass of about 1.026 kg, a forearm length of about 290 mm, a forearm mass of about 1.387 kg. The weight of the hand tool is 1 kg. The model shown in Figure 2 is created in Adams. Both the upper arm and the forearm are modeled as a uniform mass link. Fixed to the forearm, the hand tool is modeled as a 1 kg ball. The shoulder and elbow joints are considered as two rotating joints.

Figure 1. The trajectory of upper limb.

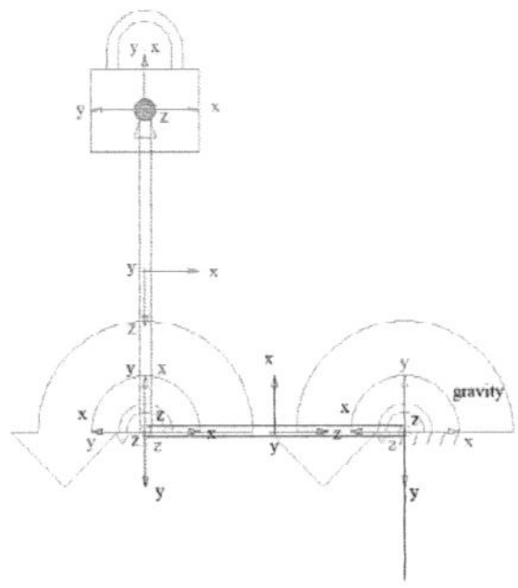

Figure 2. Dynamics model of upper limb.

According to the above model, the force moment required for the rotation angle of the shoulder joint and the elbow joint with the upper arm can be obtained as shown in Figure 3. As can be seen from the figure, the torque requirement of the shoulder joint is much larger than that of the elbow joint. The torque curve of the shoulder joint is in the form of a parabola, which increases first and then decreases. The maximum torque occurs when the angle of rotation is approximately 30 degrees and the maximum value is 10564 N· mm.

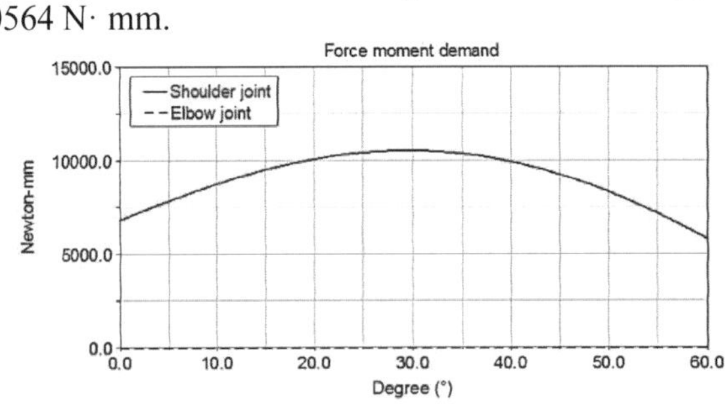

Figure 3. The force moment required for the rotation angle of the shoulder joint and the elbow joint with the upper arm.

4. Joint drive design of upper limb exoskeleton

The shoulder joint is mainly stressed during the assembly work of the worker, so it is necessary to design the support torque as close as possible for the shoulder joint. As shown in Figure 4, the upper limb exoskeleton provides a support plate that is bound to the upper arm to serve as a lifting arm and a rotational joint is designed to provide sufficient torque at the shoulder joint.

Since the torque curve of the shoulder joint is in the form of a parabola and the torque provided by the rotating joint should be as close as possible to this curve, the internal structure of the rotating joint is designed as a crank-link mechanism, as shown in Figure 5. The brace is fixed at an angle to the crank. When the crank rotates around the shell, the piston slides down the channel of shell and compresses the spring. The spring force provides a moment for the rotating joint.

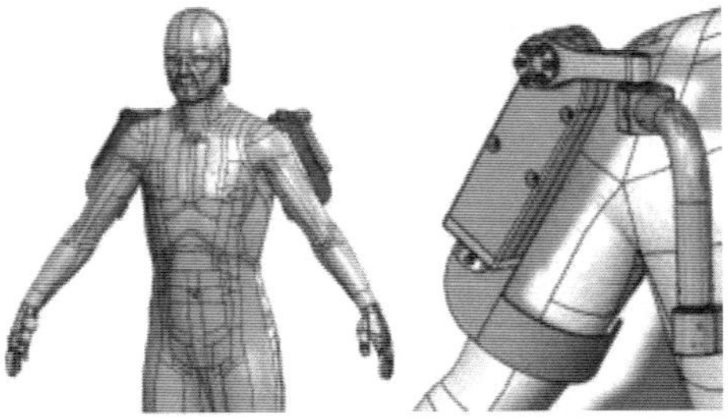

Figure 4. Joint drive design of upper limb exoskeleton.

5. Simulation and optimization of joint driving force effect

5.1. Simulation and optimization of joint driving force effect

According to the internal structure of the rotating joint and the trajectory of the arm, an equivalent mechanical analysis model can be established on Adams, as shown in Figure 6. Based on the model of Figure 2, a crank-link mechanism is added, the crank is fixed at 355° to the upper arm. The crank length is 8 mm, and the connecting rod length is 20 mm. The friction of each rotating joint is ignored. When the size of the exoskeleton element is clear, the joint driving force effect is mainly related to the spring stiffness and the pre-compression force.

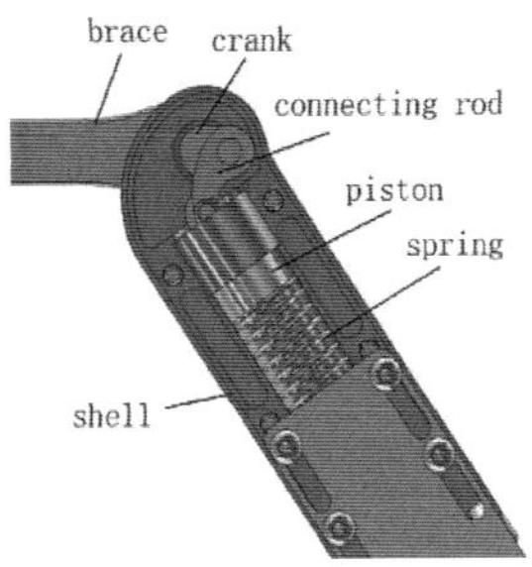

Figure 5. The internal structure of the rotating joint.

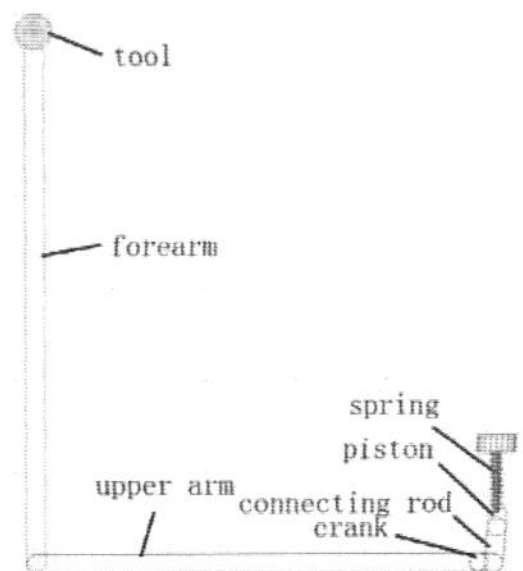

Figure 6. The equivalent mechanical analysis model.

The effect of the pre-compression force on the joint driving force is shown in Figure 7. When the spring stiffness is 136 N/mm and the pre-compression force is 750 N to 950 N, The additional required force moment for the shoulder joint to rotate with the upper arm is shown in Figure 7.

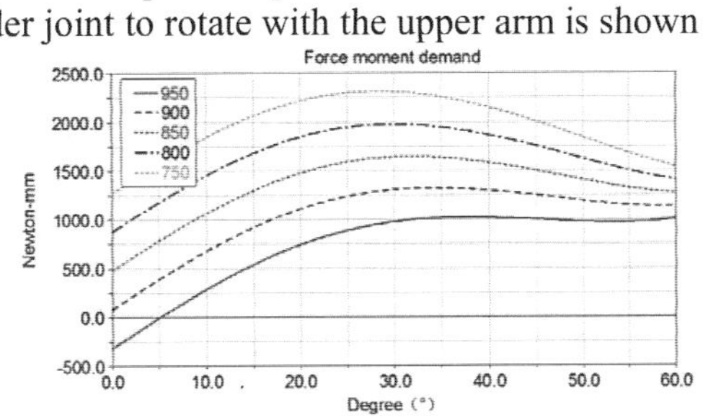

Figure 7. The effect of the pre-compression force on joint driving force.

From Figure 7, it is concluded that after wearing the exoskeleton, the maximum torque required for the shoulder joint drops below 2500 N· mm, which greatly relieve the fatigue of the shoulder. At the same time, under this setting condition, as the pre-compression force increases, the effect of the exoskeleton is better, and the initial additional required force moment of the shoulder joint is reduced.

However, when the pre-compression force reaches 950 N, a negative initial additional required force moment appears which indicates that the arm has a tendency to rotate in the direction of anti-gravity, and the design should avoid this state.

The effect of the spring stiffness on the joint driving force is shown in Figure 8. When the pre-compression force is 800 N and the spring stiffness is 100 N/mm to 140 N/mm, the additional required force moment for the shoulder joint to rotate with the upper arm is shown in Figure 8.

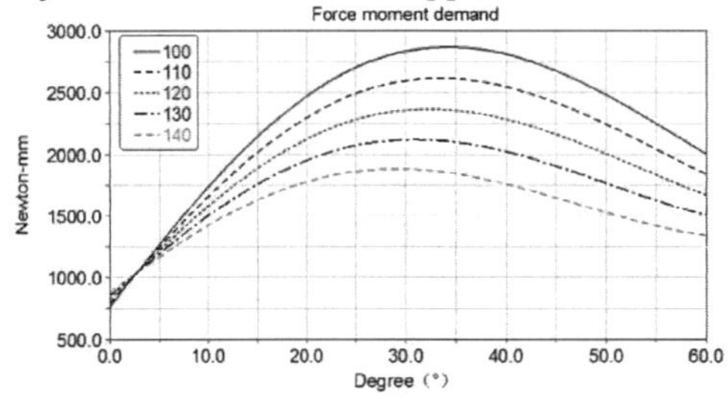

Figure 8. The effect of the spring stiffness on joint driving force.

Figure 8 shows that under this setting condition, as the spring stiffness increases, the exoskeleton boosting effect is better, and the additional required force moment curve of the shoulder joint is also more gradual.

5.2. Optimization of joint driving force effect

Pre-compression force and spring stiffness have a great influence on the driving effect. To derive the most appropriate design variables, an objective function needs to be established for this boosting effect. When the upper arm rotation angle is set to φ, the additional required force moment of the shoulder joint should be $f(\varphi)$. According to the actual spring stiffness K and the pre-tightening force F_0, the work of the additional required force moment is set as the objective function and the constraint conditions are as follows.

$$\begin{cases} min\left(\frac{\pi}{180}\int|f(\varphi)|\,d\varphi\right) \\ 11 \le K \le 153 \\ 0 \le F_0 \le 900 \end{cases} \tag{1}$$

Under the condition of setting the above objective function and design variable on Adams, the optimization result is shown in Figure 9. The maximum torque is only 912 N· mm. The optimization results show that the spring stiffness K is 153 N/mm, the pre-compression force F_0 is 900 N, and the work of the additional required force moment is 769.5 J. However, in the absence of exoskeleton assistance, the work should be 9503.3 J. The exoskeleton can save about 90% of energy consumption.

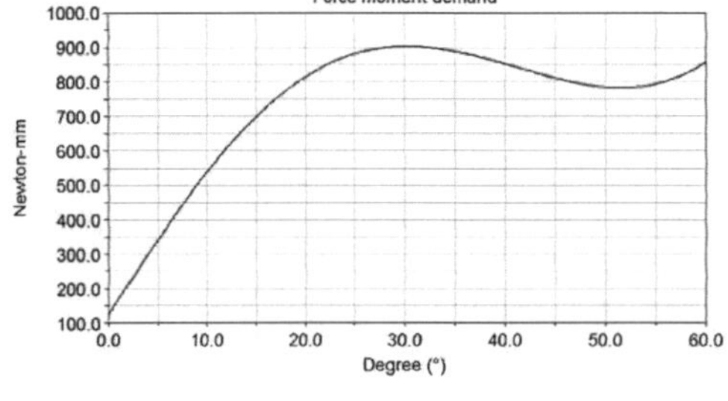

Figure 9. The optimization of the additional required force moment.

6. Conclusion

A wearable upper limb exoskeleton for factory assembly positions was designed. The trajectory and dynamics of the worker's arm during assembly was analyzed. According to the power demand, the corresponding torque drive scheme at the shoulder joint was designed. The effect on joint drive force was simulated on Adams. As shown by the result, pre-compression force and spring stiffness have a great influence on the driving effect. Finally, with the work of additional required force moment for the shoulder joint as the objective function, the two parameters of spring stiffness and pre-compression force were optimized. Compared to not wearing the exoskeleton, the energy consumption of the exoskeleton was reduced by 90%.

The research in this paper suggests that the use of the wearable upper limb exoskeleton seems to be effective in helping workers save energy consumption during long hours of work, which is consistent with the conclusions of many scholars on upper extremity exoskeleton studies[11, 12]. Since occupational injuries of the upper limbs involved in the assembly process cannot be completely eliminated in the workplace [13], it is of great practical significance to study auxiliary equipment such as upper limb exoskeleton. We need further research and improvement of the design of the upper limb exoskeleton in the future, and evaluate the long-term effects of using upper limb exoskeleton, the effects on different populations and the potential benefits in actual workplaces (eg cars, aircraft assembly lines).

7. References

[1] Aliman N, Ramli R and Haris S M 2017 Design and development of lower limb exoskeletons: A survey *Robotics and Autonomous Systems* **95** 102-16

[2] de Kruif B J, Schmidhauser E, Stadler K S and O'Sullivan L W 2017 Simulation Architecture for Modelling Interaction Between User and Elbow-articulated Exoskeleton *Journal of Bionic Engineering* **14** 706-15

[3] Galle S, Derave W, Bossuyt F, Calders P, Malcolm P and De Clercq D 2017 Exoskeleton plantarflexion assistance for elderly *Gait & Posture* **52** 183-8

[4] Hansen C, Gosselin F, Ben Mansour K, Devos P and Marin F 2018 Design-validation of a hand exoskeleton using musculoskeletal modeling *Applied Ergonomics* **68** 283-8

[5] Anam K and Al-Jumaily A A 2012 Active Exoskeleton Control Systems: State of the Art *Procedia Engineering* **41** 988-94

[6] Singla A, Rupal B S and Virk G S 2016 Optimization of stepped-cone CVT for lower-limb exoskeletons *Perspectives in Science* **8** 592-5

[7] Ko H K, Lee S W, Koo D H, Lee I and Hyun D J 2018 Waist-assistive exoskeleton powered by a singular actuation mechanism for prevention of back-injury *Robotics and Autonomous Systems* **107** 1-9

[8] Stewart A M, Pretty C G, Adams M and Chen X 2017 Review of Upper Limb Hybrid Exoskeletons *IFAC-PapersOnLine* **50** 15169-78

[9] Dahmen C, Wllecke F and Constantinescu C 2018 Challenges and Possible Solutions for Enhancing the Workplaces of the Future by Integrating Smart and Adaptive Exoskeletons *Procedia CIRP* **67** 268-73

[10] Riani A, Madani T, Benallegue A and Djouani K 2018 Adaptive integral terminal sliding mode control for upper-limb rehabilitation exoskeleton *Control Engineering Practice* **75** 108-17

[11] Rashedi, E., et al., Ergonomic evaluation of a wearable assistive device for overhead work. Ergonomics, 2014. **57**(12): p. 1864-1874.

[12] Kadota, K., et al., Development of power-assist robot arm using pneumatic rubber muscles with a balloon sensor. The 18th IEEE International Symposium on Robot and Human Interactive Communication, 2009: p. 546-551.

[13] Kim, S., et al., Assessing the influence of a passive, upper extremity exoskeletal vest for tasks requiring arm elevation: part i – "expected" effects on discomfort, shoulder muscle activity, and work task performance. Applied Ergonomics, 2018.

Acknowledgements

This paper is supported by Guangzhou Scientific and Technological Projects (NO.201802010067).

IOP Conf. Series: Materials Science and Engineering **717** (2020) 012005 doi:10.1088/1757-899X/717/1/012005

Identification and Control of Counter-traction Mechanism to Finger Traps for Fracture Reduction

N Prasoetkarn[1], N Numanoy[1], S Saengsri[1] and J Srisertpol[1]

[1] School of Mechanical Engineering, Institute of Engineering, Suranaree University of Technology, Nakhon Ratchasima 30000, Thailand

E-mail: jiraphon@sut.ac.th

Abstract. Finger traps are a valuable tool that can be maintained the traction. The devices stretch the thumb and other four fingers separately in order to achieve better reduction effect. These are applied individually to the fingers and the limb is suspended, with gravity providing counter-traction to disimpact the fracture by traction. However, personal gravity can't be used to maintain the appropriate tension. This paper proposed an experimental study on the tension of counter-traction mechanism to finger traps for fracture reduction. This mechanism is particularly useful to help for the patients. Transfer function estimation from experimental data can improve the overall identification performance and the parameter. The feedback control system design for compensating tension that loses from disturbance tension consisted of the controller as PI controller and actuator as DC motor with counter-traction mechanism. Experimental results were demonstrated the efficiency of the proposed system to maintain the appropriate tension.

1. Introduction

Recent decades, surgical approaches such as open reduction and internal fixation have been increasing use. Complications such as tendon afflictions and further surgery can arise from surgical intervention. Therefore, closed reduction and cast immobilization remains an important treatment option in a majority of cases. A very commonly used method of closed reduction is manual traction. An assistant provides counter-traction while the operator provides traction and manipulates the bone fragments into position. Even though it has been the most commonly used method for at least the better part of a century, the evidence of its effectiveness is ambiguous [1]. Another method is mechanical reduction by finger-trap traction [2] which dispenses with the need for an assistant as the forearm is suspended by finger-traps in the radial fingers. The traction mechanism stretches the thumb and other four fingers separately in order to achieve better reduction effect [3]. Counter-traction is provided by weights suspended on the arm near the elbow joint with gravity as shown in Figure 1. This restores the longitudinal axis without further actions, and the operator can then apply manual dorsal pressure to the fragments, if necessary, to restore the volar tilt of the wrist [4]. Closed reduction by finger-trap traction seems to offer better correction of radial shortening. However, the optimal weight to relate with tension of closed reduction remains to be determined.

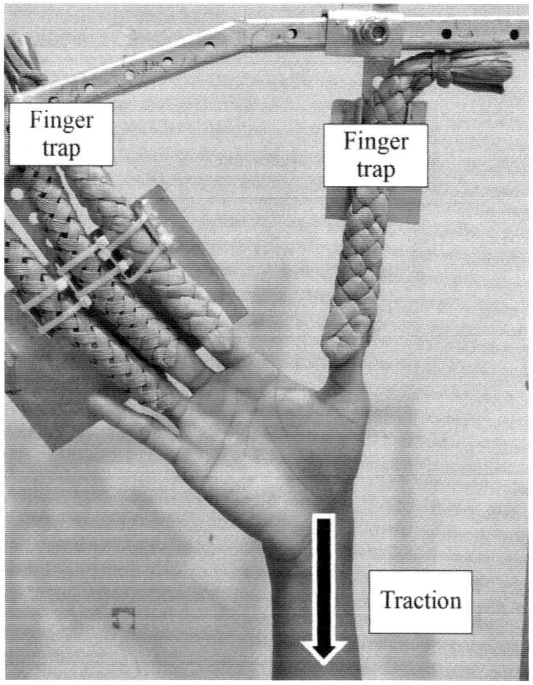

Figure 1. Finger-trap traction

This paper proposed an experimental study on the tension of counter-traction mechanism to finger traps for fracture reduction. This mechanism is particularly useful if there is no assistant available. Simplistic transfer function, to improve the overall identification performance, the parameters and mechanism were first estimated using a DC motor signal. The PI controller with the DC motor control has been design based on the concept of feedback control system for compensating tension load that loses from disturbance tension.

2. System modelling

The objective is to model the interaction between the counter-traction mechanisms with tensions of DC motor control. Although the counter-traction mechanism that controls the positions and torques of mechanism in follow the tensions generated by controller. In general, the drive system contains two major parts. The first one is the electronic control system, actuator, controller, and load cell. The second one is the mechanical system such as the motor, lead screw, spring, and tension connector, etc. The control system with feedback is a closed-loop system as shown in Figure 2.

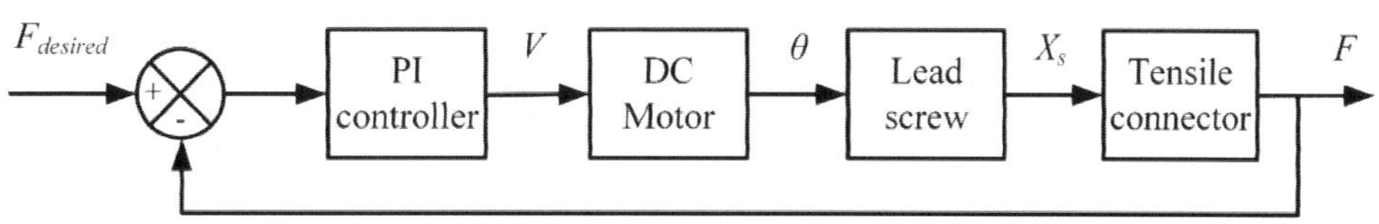

Figure 2. The diagram of counter-traction mechanism

2.1. Counter-traction mechanism

The motor models have been proposed for system identification and controller design [5]. The motor torque is transmitted to the lead screw with tension connector that has a damping coefficient of rotary motion (B) and angular displacement (θ)

$$T = J\frac{d^2\theta}{dt^2} + B\frac{d\theta}{dt} + T_L \tag{1}$$

where *J* is the total inertia of coupling, motor shaft, tension connector and lead screw.
The torque, which is used to move the tension connector, is generated by applying input voltage (V) to the amplifier which has a torque constant (K_t), such as

$$V = L\frac{di}{dt} + iR + K_e\frac{d\theta}{dt} \tag{2}$$

$$T = K_t i \tag{3}$$

The current (i) is fed to control the DC motor, where L, R and K_e are inductance, resistance, and back electromotive force constant of DC motor, respectively. The tension load (F) from wire tension at nut of lead screw is transmitted as the axial load torque (T_L) as

$$T_L = F \otimes r \tag{4}$$

Where $\otimes$ denote the cross product operation is perpendicular. The total axial displacement of the tension connector is contributed by the axial movement of the lead screw reflected at the nut (X_s) and the angular displacement (θ) of the lead screw translated as the axial displacement [6].

$$F = K_s X_s = K_s r\theta \tag{5}$$

where K_s is tension gain of spring, and the transmission ratio is $r = p/2\pi$ with pitch length (p).
Taking Laplace transform of equations (1) through (5), the transfer function between the tension load (F) and input voltage (V) can be expressed as

$$\frac{F}{V} = \frac{rK_tK_s}{JLs^3 + (BL + JR)s^2 + (r^2LK_s + BR + K_tK_e)s + r^2RK_s} \tag{6}$$

The mathematical model in equation (6) is the theoretical analysis of dynamic phenomena of the counter-traction mechanism. The parameters of the model aren't an unknown and uncertain values.

2.2. System identification

Methods of system identification, parameter estimation and optimization applied to problems of modelling and control. The counter-traction mechanism is the closed-loop control process [5]. The mathematical model can be estimated from experimental data. In this, the most popular method for curve-fitting is described. Collection data base on time domain, the desired tension was constant at 3 N, an initial condition. The input signal was generated the desired reference a force between 50 to 80 N and each of the step was changed 10 N with 400 seconds.

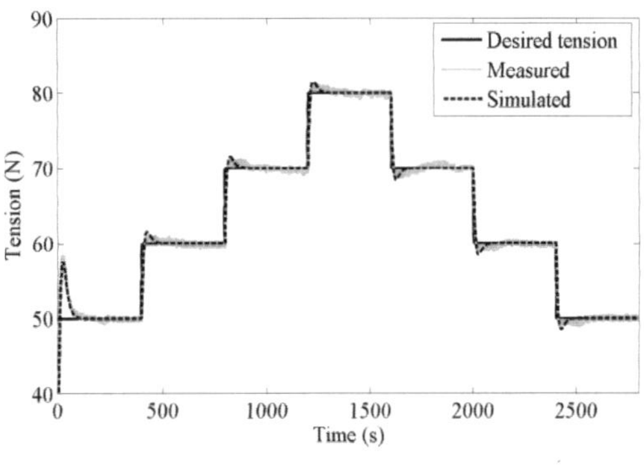

Figure 3. The performance of identification and measurement

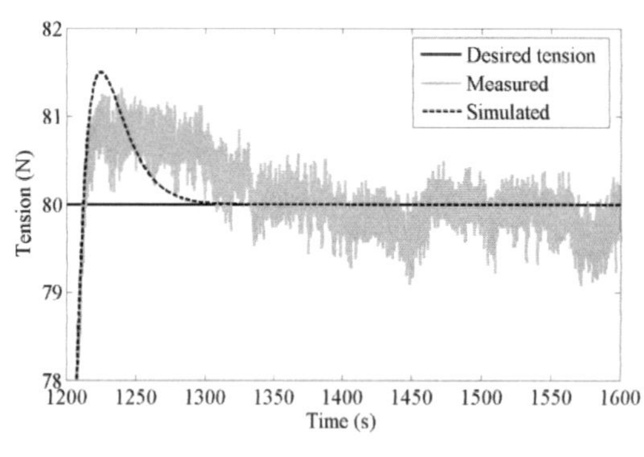

Figure 4. Comparative convergence of the response at 80 N

Thus, the approximated parameters from identifications as shown in table 1, and rearrangement are given by

$$\frac{F}{V}=\frac{b_0}{s^3+a_2s^2+a_1s+a_0} \tag{7}$$

Table 1. Parameters of counter-traction mechanism

Symbols	Values
$b_0 = rK_tK_s/JL$	2.6
$a_0 = (BL+JR)/JL$	4.4e-3
$a_1 = (r^2LK_s+BR+K_tK_e)/JL$	9.6e-1
$a_2 = r^2RK_s/JL$	9.7e-1

The responses that behaviors of the system identification to simulate (solid line), measurement (gray line), and desired tension (dashed line) were shown in figures 3 and 4. In the extended scale in a point of tension at 80 N, the behavior transients of the convergence comparisons can be expressed the response as shown in figure 4. The phenomenal response reveals the rise time of 8 seconds, the settling time of 9 seconds (steady state of 2%) and the overshoot of 1.5%.

3. Experimental and simulation results

In this section, to test effectiveness of the proposed control method, simulations and experiments were carried out on the counter-traction mechanism on tensions of DC motor for position control with PI controller as shown in figure 5. To realize the proposed digital control system using a digital signal processor (DSP), the DSP-board manufactured by Maneesoon Group Company Limited (MNS) was utilized. Control algorithms were programmed in the MATLAB and Simulink environment and were then compiled and loaded to the DSP-board. The sampling frequency was 1 kHz. At each sample time, four successive 12-bit A/D conversions of each signal were taken. Moreover, there were interface circuits for command and data communication between the computer and the development module. In addition, the corresponding tension load was measured via load cell sensor. The input range of each channel was from 0 to 5 volts.

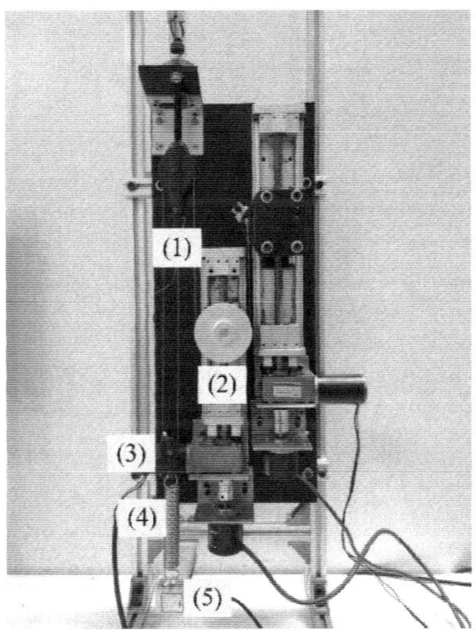

Figure 5. The counter-traction mechanism, (1) tension connector, (2) lead screw, (3) DC motor, (4) spring, and (5) load cell sensor

3.1. Performance of desired input tracking

In the response optimization of PI controller, the maximum overshoots by 2% in the step response. The rise time and settling time are less than 15 (80%) and 20 (2%) seconds. The controller parameters are optimized to be P = 0.1219 and I = 7.8097e-04.

For comparison purpose, a display the response impulsive reference using input tracking based on optimal controller, shown in figures 6 and 7. These results strongly suggest the optimal controller that were capable reducible the effect of the overshoot and steady-state error.

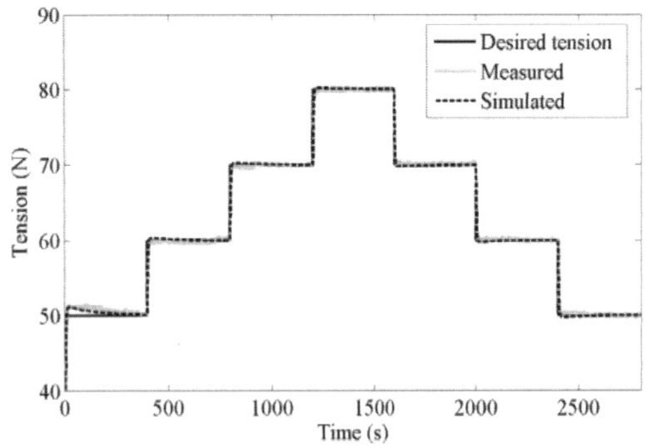

Figure 6. The tracking performance of optimal PI controller

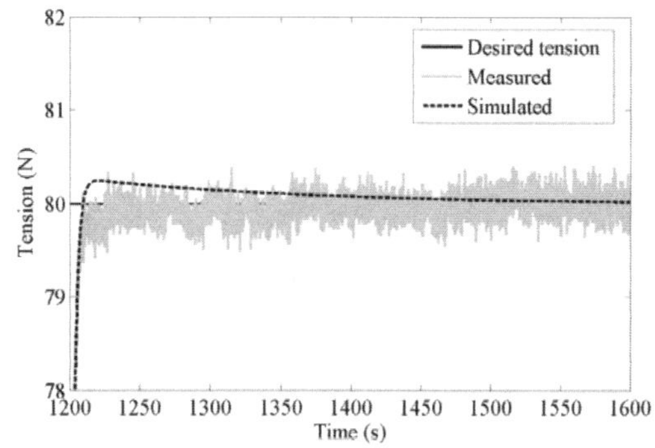

Figure 7. Comparative convergence of the response at 80 N with optimal PI controller

3.2. Robust disturbance

In the final test, the experimental result with a disturbance response was studied with the response effect to the dynamics of the system when there is a change of the external force interference. Applying traction to fracture reduction requires constant tension. For disturbance testing, changing the position of tension connector upward or downward one step will change the tension increase or decrease with magnitude of 8 N. There are two types of controller's working conditions.

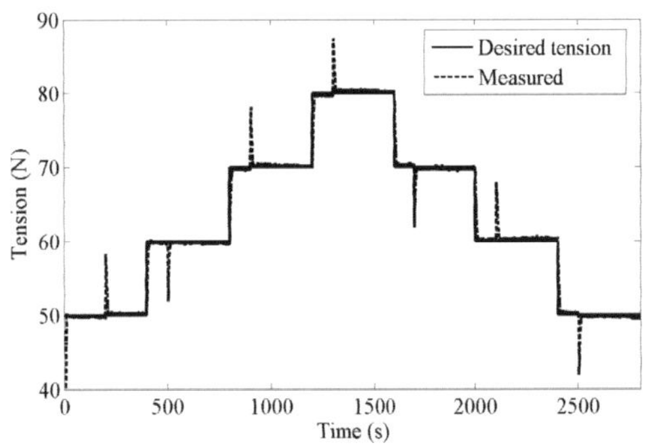

Figure 8. The responding of immediate self-adjustment

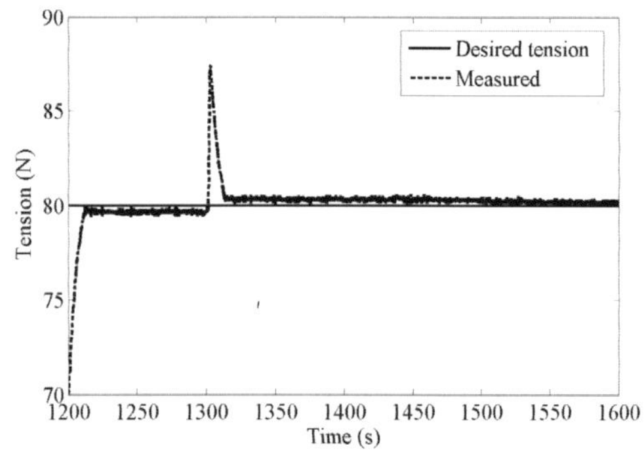

Figure 9. The transient response of immediate self-adjustment at 80 N

3.2.1 Immediate self-adjustment. The disturbance of response is a function with a peak amplitude from the set point. This is considered the robust external interference that occurred. When the system was disturbed, the output tension changed from desired tension that controller could be adjusted tension by itself. Figures 8 and 9 show the response of the optimal PI controller with the disturbance. At steady state 80 N, the system was disturbed by increasing 8 N at time of 1300 sec. The maximum tension is 88 N. After that the system can be recovered to steady state with the settling time of 10 seconds (2%). The controllers are suddenly robust to external disturbance, but the response occurred the overshoot more than 10%

3.2.2 Conditional self-adjustment. The actual application, the disturbances from the tension can be predicted the boundary. So, the system was disturbed with 8 N, then an increase or decrease disturbance occurred the boundary at ±5% of the references. The controllers are suddenly robust to external disturbance over ±5% of the references and aren't adjusted the tension under boundary as shown in figures 10 and 11, respectively. The controllers are suddenly robust to external disturbance in the boundary conditions.

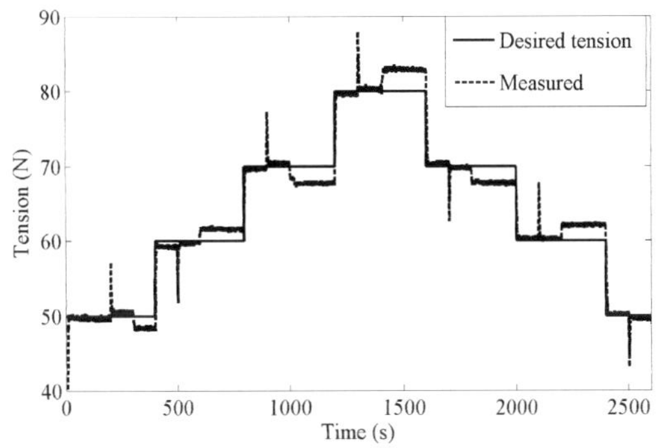

Figure 10. The responding of conditional self-adjustment

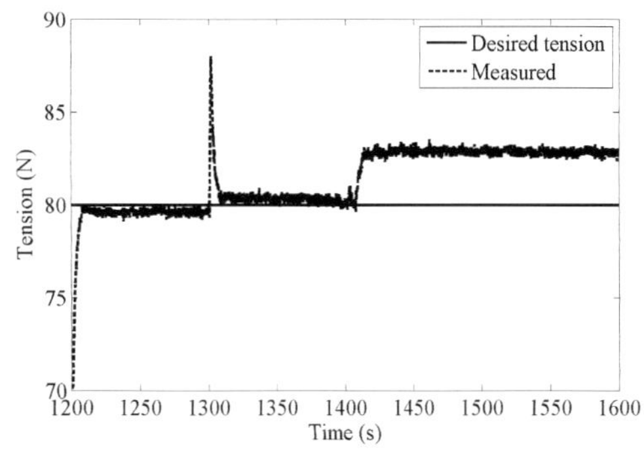

Figure 11. The transient response of conditional self-adjustment

4. Conclusions

The analytical and experimental results of counter-traction mechanism demonstrate the effectiveness of identification and control of wire tension in a closed loop system. In the beginning, prescribed the PI controller for the system are applied to maintain position with relative the tension load. Transfer function of estimation explained, the phenomenon between tension and controller was designed by using an optimal PI controller. The results of a comparative study between an immediate and conditional self-adjustment for the counter-traction mechanism were discussed in terms of transient responses and steady state responses. The feedback system can be maintained the tension with compensation that loses from disturbance in specific range of 50-80 N.

5. References

[1] Charnley J 1961 *The closed treatment of common fractures* (New York: John Charnley) p 128
[2] Holkenborg J, Napel S -J T and Kolkman K 2013 *Annals of Emergency Medicine* **62** 4 S66
[3] Del Pinal F 2011 *American Society for Surgery of the Hand* **36** 10 1694
[4] Xie Z, Liang B, Wang X and Tian M 2016 *Proc. 23rd Int. Conf. on Mechatronics and Machine vision in Practice (Nanjing)* (New Jersey: IEEE)
[5] Khajorntraidet C, Srisertpol J and Prattanarak C 2013 *Advanced Science Letter* **19** 8 2139
[6] Lin C Y and Lee C H 2017 *Applied science* **7** 8 776

Acknowledgments
The authors are greatly indebted to Suranaree University of Technology (SUT) and the Maneesoon Group Company Limited (MNS), for their generous support and their valuable comments.

IOP Conf. Series: Materials Science and Engineering **717** (2020) 012006 doi:10.1088/1757-899X/717/1/012006

Advances in Precision Control of Magnetic Microrobots

Aalim M. Mustafa [1,*] **, Nassim Khaled** [1]

[1] Mechanical Engineering Department, Prince Mohammad bin Fahd University, Khobar, KSA.

[*] corresponding author's email: amustafa1@pmu.edu.sa

Abstract. Magnetic actuated systems promise great commercial applications. For instance, it can be used to precisely control the targeted drug delivery and minimally invasive medical applications. In this paper, we have summarized recent advances in the field of control and path planning of magnetic microrobots. We have investigated the main differences between approaches, materials, trajectories, control methods and the main features of each research paper.

1. Introduction

The study of magnetic microrobots dates back to 3 decades [1]. The research gained special attention due to the potential usage in drug delivery industry [2-9] as well as medical applications such as MRI-scanners [10-16]. Multiple publications documented the progress of magnetic microrobots [17-22]. Nevertheless, there is no recent review focused on development of control strategies applied in microrobots for path following. Figure 1 describes the progress of magnetic microrobots experimental research in the last ten years based on our data. It shows the number of publications including the phrase "magnetic microrobot" in its title. The statistics we got is closely matching an analysis by ISI web of knowledge©, reported in March 2019, when searched by the same phrase.

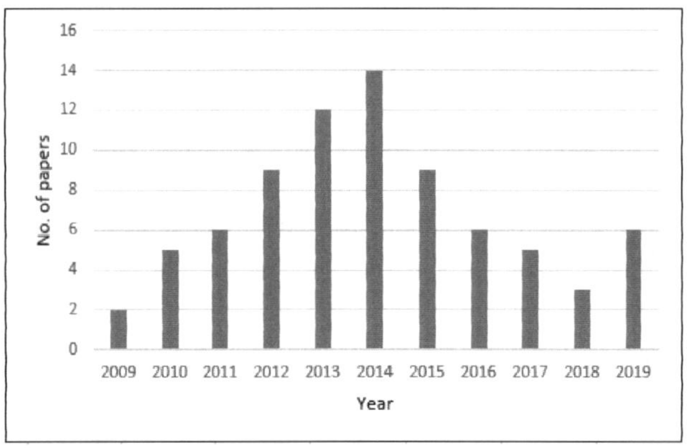

Figure 1. Distribution of scientific papers (reported in this review) in the last ten years.

In this survey, we will concentrate on research developed on experimental setups for magnetic microrobots controlling and tracking. The paper is organized as follows: section II provides a background of magnetic actuation and past contributions in the topic. Section III presents information on magnetic robot's specifications, size, geometry, material, magnetic device configuration and the working medium. Section IV offers samples of trajectories designed in literature for magnetized microrobots to follow. Section V discusses the control methods applied to stabilize the magnetic robots, reject disturbances and follow the desired trajectory. Section VI covers the conclusion and provides recommendations for future studies.

2. Background

Magnetic field is a popular actuation method for microrobots due to its high actuation force, compact system size, and low hardware cost. Microrobots are made of permanent magnetic, ferromagnetic, or paramagnetic material. Any magnetized robot will experience a torque and a force due to the electromagnetic field that can be described as follows:

$$\boldsymbol{F_m} = V_r(\boldsymbol{M}.\nabla)\boldsymbol{B}(x,y,z) \quad (1)$$

$$\boldsymbol{T_m} = V_r\, \boldsymbol{M} \times \boldsymbol{B}(x,y,z) \quad (2)$$

F_m and T_m are the force and torque experienced by the robot, V_r is the volume of the robot, M is the magnetization of the robot, and B is the magnetic potential produced by the electromagnetic field. By controlling the magnetic field, the dynamics of the microrobot can be controlled to realize autonomous navigation. Significant body of simulation and experimental research work have been invested in the field [23]. Navigation in high viscous fluids under pulsatile flow was studied [24]. Additionally, controlling multi-agent independent microrobots [15,25-28] who share a global driving magnetic signal is another complex field that has been investigated.

3. Microrobot Specifications

By far the most reported geometry for the microrobots is the sphere It has a diameter range between $20 - 600\ \mu m$. Additionally, cuboid and cylindrical microrobot were reported. Table 1 shows geometry, dimensions, and materials of the microrobots. It is worthwhile noting that all dimensions listed in table 1 are in micrometer. The sole article we found that reported millimeter size robots is article [29] (almost 50 times bigger than a typical microrobot).

Table 1. Material, geometry, and dimensions of the microrobots

Ref.	material	Shape	dimension	Ref.	material	shape	dimension
[24]	neodymium-iron-boron	spherical	$r = 250\ \mu m$	[34]		spherical	$r = 50\ \mu m$
[30]	NdFeB N35	spherical	$r = 257\ \mu m$	[35]	Polymer (SU-8)+CoNi	cuboid	$50 \times 60 \times 300\ \mu m$
[31]	Fe3O4	spherical	$r = 100\ \mu m$	[36]	neodymium-iron-boron	spherical	$r = 250 - 300\ \mu m$
[32]	Fe3O4	spherical	$r = 15\ \mu m$	[37]		spherical	$r = 10\ \mu m$
[30]	NdFeB	cylindrical	$r = 15.87\ mm$ $h = 31.75\ mm$	[28]	Polymer (BJB M-3184)	spherical	$r = 250\ \mu m$
[33]	NdFeB+PDMS	cylindrical	$r = 300\ \mu m$ $h = 200\ \mu m$				

IOP Conf. Series: Materials Science and Engineering **717** (2020) 012006 doi:10.1088/1757-899X/717/1/012006

The position of the electromagnets of a system in space has a large influence on the reachable magnetic fields and gradients in certain orientations [38]. Additionally, the medium surrounding the microrobot has an effect on systems performance. Researchers tried to use interfaces and solutions with properties closed to the blood viscosity and density, since most suggested applications of magnetic microrobots are in the medical field. Table 2 shows some water solutions used as a working medium.

Table 2. Working medium of the microrobots

Ref.	Coils configuration	Medium	Magnetic field strength	Ref.	Coils configuration	Medium	Magnetic field strength
[24]	3 Maxwell + 1 Helmholtz	water + glycerol	300 – 350 mT	[35]	MiniMag	ultra-sonicated ethanol+water	20 mT
[30]	3 Maxwell + 1 Helmholtz	glycerine-water solution	300 mT	[36]	3 Maxwell + 1 Helmholtz	water + glycerol	30 mT
[31]	OctoMag	poly d,L-lactic acid+water	100 mT	[28]	Helmholtz	water-oil	15 mT
[39]	5 OmniMagnet	water	1.5 mT	[40]	3 Maxwell + 1 Helmholtz	water + glycerol	40 mT

4. Trajectories

The studied literature has a wide range of selected trajectories designed for the magnetic microrobot's motion. Some authors considered the two-dimensional motion of the magnetic microrobots [41-44], while others generalized to the three-dimensional motion [45-51]. In reality, paths of the robot can be blocked, thus many researches studied online path planning and designed obstacle avoidance algorithms.

Figure 2 shows five open-loop linear trajectories [24, 29, 30-31, 33-35 and 52]. Figure 3 shows four closed-loop trajectories (30, 32 and 33]).

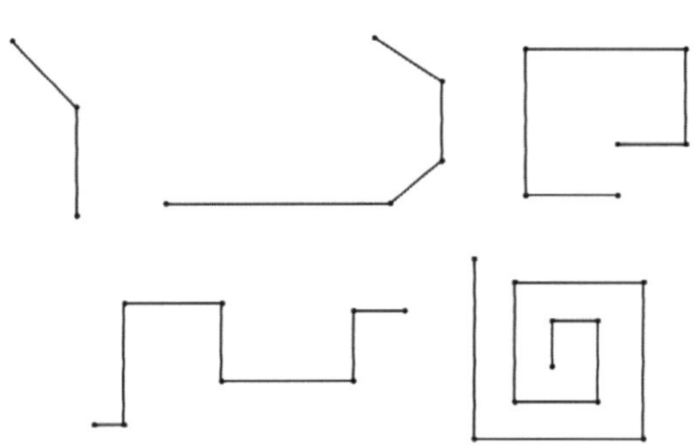

Figure 2. Open-loop linear trajectories

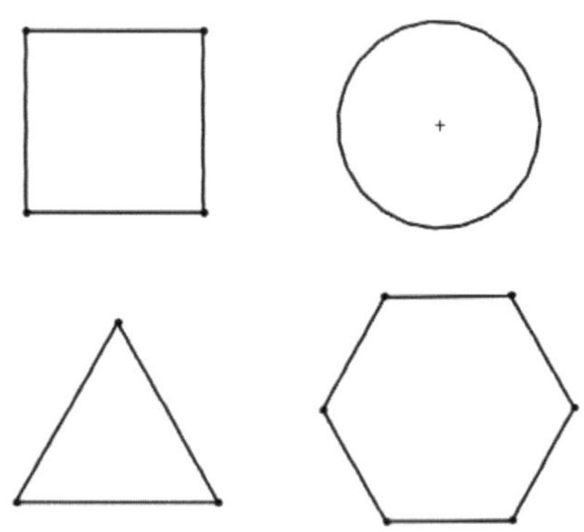

Figure 3. Closed-loop trajectories

IOP Conf. Series: Materials Science and Engineering **717** (2020) 012006 doi:10.1088/1757-899X/717/1/012006

Figure 4 shows three trajectories with combination of both linear and curved segments [53], [33], and a circular trajectory with obstacle avoiding [35]. Figure 5 shows two trajectories in three dimensions, a 3D helical trajectory [37] and a 3D cubic trajectory [39].

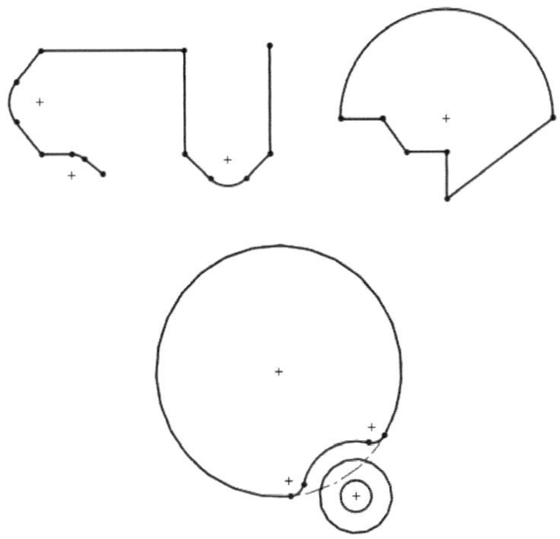

Figure 4. Linear and curved segments trajectories

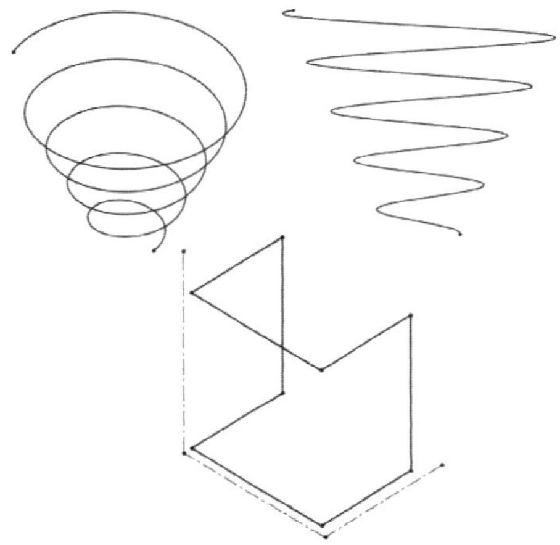

Figure 5. Three dimensions trajectories

5. Control Strategy

Open-loop controllers are sometimes used to control motion of magnetic microrobots [36]. More commonly, feedback controllers were widely used to precisely control magnetic microrobots under complex environment and subject to variability. Complexity of controlling magnetic microrobots increases exponentially with the degrees of freedom of the system [18, 54-59].

[60] applied a proportional plus integral controller to drive the magnetic field to follow a specified trajectory. [52] applied a proportional derivative controller for tracking position and velocity of the microjet. [34] applied a proportional derivative controller and a force observer for disturbance compensation of the microrobot motion. [61] used a proportional derivative (PD) controller with a Disturbance Observer (DOB) to accurately track the position command and compensate for non-predicted disturbances. They also applied Kalman filter for noise rejection. [28] applied a proportional controller to control the 2D motion of small scale (mm sized) identical magnetic microrobots in close proximity to each other. The research studied different control inputs based on the tracking method, whether to track a desired radial or transverse distances, or angels between the two reports.

[24] used generalized predictive controller (GPC). It is sufficiently robust against nonlinear model uncertainties (e.g. drag force and viscosity), external perturbations (systolic pulsatile flow) and noisy trajectory tracking measurements. In GPC scheme design the system is usually modeled using the model Controlled Auto-Regressive Integrated Moving-Average (CARIMA). [35] Designed a model predictive controller (MPC) on a linearized model, to perform the task of trajectory tracking and obstacle avoidance. [40] proposed a linear quadratic integral (LQI) controller to demonstrate the optimal stabilization of a desired positions of two magnetic microrobots navigating in a microfluidic channel.

[30] developed a control law for the nonlinear model using an adaptive back-stepping control approach, coupled with a high gain observer. [32] designed a controller based on the theory of input-to-state stability and recursive application of back-stepping to drive the micro particle to follow successfully the desired trajectory. [53] designed a model-free extended state observer (ESO) based trajectory tracking control scheme for two magnetic microrobots. [29] designed a kinematic controller

as well as a trajectory generation technique that takes the non-holonomic constraint of the robot into account. it was a model-free approach based on visual feedback of the micro object.
[31] developed a two loops control system. The inner loop estimates the disturbance force and provides a compensating control input based on the disturbance force observer, while the outer loop achieves stability of the overall magnetic system through a proportional derivative controller. [37] compared three control techniques to for trajectory tracking. First, input-to-state stability (ISS) based controller, second, high-gain observer-based controller, and compared both of them with a traditional PID controller.
[33] proposed a control algorithm to control a microrobot with an electromagnetic manipulation system. The microrobot performs a point-by-point path-tracking motion. With the control algorithm and the visual feedback, the driving force is reducing when the microrobot is nearing a target point. The tracking path was separated to several segments with target points. The moving speed was controlled to reduce to zero when the microrobot arrives at every target point.

6. Conclusion
In this this article, we have provided a categorical summary of published articles in the field of magnetic microrobots. Size of the application, trajectory type as well as control scheme were the categories we used. The different types of controllers applied in literature performed a fairly acceptable tracking, in spite of complexity of the reference trajectory. The position error was always less than half of the microrobot size.
The geometry of microrobots under study was mostly spherical, with a maximum radius of 300 μm. Manipulation of this very small size needed a magnetic field strength around 300 mT, which indicates the proportional relation between size of the microrobot and the strength of the applied magnetic field. Scaling the robots poses a challenge related to magnetic coil limitations. When coil dimension increases, it will be subjected to an excessive heating, which requires a proper cooling system.
In order to simulate the magnetic microrobots applications, the workspace of most mentioned articles was manipulated in liquid environments. These have viscosity and density similar to blood. In liquid mediums, drag force reduces the velocity of the magnetic microrobots, therefore, attention should be taken to overcome this loss of speed.
Based on success and development of magnetic microrobots, it has a great potential for many medical, biological and industrial applications. They have a very small sizes, low weight, and they are able to perform tasks accurately in complex environments. Scaling them to bigger applications is still a challenge.

7. References
[1] T. Honda, K. I. Arai, and K. Ishiyama, “Micro Swimming Mechanisms Propelled by External Magnetic Fields,” *IEEE Trans. Magn.*, vol. 12, no. 5, pp. 5085–5087, 1996.
[2] S. T. Ã, F. Mishima, S. Fujimoto, Y. Izumi, and S. Nishijima, “Development of magnetically targeted drug delivery system using superconducting magnet,” *J. Magn. Magn. Mater.*, vol. 311, pp. 367–371, 2007.
[3] M. S. Sakar, E. B. Steager, A. Cowley, V. Kumar, and G. J. Pappas, “Wireless Manipulation of Single Cells using Magnetic Microtransporters,” *2011 IEEE Int. Conf. Robot. Autom.*, pp. 2668–2673, 2011.
[4] I. S. M. Khalil, M. P. Pichel, L. Abelmann, and S. Misra, “Closed-loop control of magnetotactic bacteria,” *Int. J. Rob. Res.*, vol. 32, no. 6, pp. 637–649, 2013.
[5] I. S. M. Khalil, V. Magdanz, S. Sanchez, O. G. Schmidt, L. Abelmann, and S. Misra, “Magnetic Control of Potential Microrobotic Drug Delivery Systems : Nanoparticles , Magnetotactic Bacteria and Self-Propelled Microjets,” *35th Annu. Int. Conf. IEEE Eng. Med.*

Biol. Soc., pp. 5299–5302, 2013.

[6] M. Sitti, H. Ceylan, W. Hu, J. Giltinan, M. Turan, Y. Sehyuk, and E. Diller, "Biomedical Applications of Untethered Mobile Milli/Microrobots," *Proc. IEEE*, vol. 103, pp. 205–224, 2015.

[7] S. M. Jeon, G. H. Jang, H. C. Choi, S. H. Park, and J. O. Park, "Utilization of Magnetic Gradients in a Magnetic Navigation System for the Translational Motion of a Micro-Robot in Human Blood Vessels," *IEEE Trans. Magn.*, vol. 47, no. 10, pp. 2403–2406, 2011.

[8] P. Vartholomeos and C. Mavroidis, "In Silico Studies of Magnetic Microparticle Aggregations in Fluid Environments for MRI-Guided Drug Delivery," *IEEE Trans. Biomed. Eng.*, vol. 59, no. 1, pp. 3028–3038, 2012.

[9] C. Chen, L. Chen, P. Wang, L. Wu, and T. Song, "Steering of magnetotactic bacterial microrobots by focusing magnetic field for targeted pathogen killing," *J. Magn. Magn. Mater.*, vol. 479, no. 6, pp. 74–83, 2019.

[10] J. Mathieu, G. Beaudoin, and S. Martel, "Method of Propulsion of a Ferromagnetic Core in the Cardiovascular System Through Magnetic Gradients Generated by an MRI System," *IEEE Trans. Biomed. Eng.*, vol. 53, no. 2, pp. 292–299, 2006.

[11] L. Arcese, M. Fruchard, F. Beyeler, A. Ferreira, and B. J. Nelson, "Adaptive backstepping and MEMS force sensor for an MRI-guided microrobot in the vasculature," *2011 IEEE Int. Conf. Robot. Autom.*, pp. 4121–4126, 2011.

[12] J. Mathieu and S. Martel, "Steering of Aggregating Magnetic Microparticles Using Propulsion Gradients Coils in an MRI Scanner," *Magn. Reson. Med.*, vol. 1345, pp. 1336–1345, 2010.

[13] J. Mathieu and S. Martel, "Magnetic Steering of Iron Oxide Microparticles Using Propulsion Gradient Coils in MRI," *2006 Int. Conf. IEEE Eng. Med. Biol. Soc.*, pp. 472–475, 2006.

[14] K. Belharet, D. Folio, and A. Ferreira, "Endovascular Navigation of a Ferromagnetic Microrobot Using MRI-based Predictive Control," *2010 IEEE/RSJ Int. Conf. Intell. Robot. Syst.*, pp. 2804–2809, 2010.

[15] A. Eqtami, O. Felfoul, and P. E. Dupont, "MRI-powered Closed-loop Control for Multiple Magnetic Capsules," *2014 IEEE/RSJ Int. Conf. Intell. Robot. Syst. (IROS 2014)*, no. Iros, pp. 3536–3542, 2014.

[16] S. Tamaz, R. Gourdeau, A. Chanu, and J. Mathieu, "Real-Time MRI-Based Control of a Ferromagnetic Core for Endovascular Navigation," *IEEE Trans. Biomed. Eng.*, vol. 55, no. 7, pp. 1854–1863, 2008.

[17] S. Chowdhury, W. Jing, and D. J. Cappelleri, "Controlling multiple microrobots : recent progress and future challenges," *J. Micro-Bio Robot.*, pp. 1–11, 2015.

[18] S. Bouchebout, A. Bolopion, and J. A. S. Régnier, "An overview of multiple DoF magnetic actuated micro-robots," *J. Micro-Nano Mechatronics*, pp. 97–113, 2012.

[19] M. Arruebo, R. Fernández-pacheco, M. R. Ibarra, and J. Santamaría, "Magnetic nanoparticles for drug delivery," *Drug Dev. Res.*, vol. 2, no. 3, pp. 22–32, 2007.

[20] B. J. Nelson, I. K. Kaliakatsos, and J. J. Abbott, "Microrobots for Minimally Invasive Medicine," *Annu. Rev. Biomed. Eng.*, pp. 55–85, 2010.

[21] T. Xu, J. Yu, X. Yan, H. Choi, and L. Zhang, "Magnetic Actuation Based Motion Control for Microrobots : An Overview," *Micromachines*, pp. 1346–1364, 2015.

[22] Y. Jia, Z. Chuang, W. Xiao-dong, W. Wenxue, X. Ning, and L. Lian-Qing, "Development of micro- and nanorobotics : A review," *Sci. CHINA Technol. Sci.*, vol. 62, no. 1, pp. 1–20, 2019.

[23] L. Sadelli, M. Fruchard, and A. Ferreira, "2D Observer-Based Control of a Vascular Microrobot," *IEEE Trans. Automat. Contr.*, vol. 62, no. 5, pp. 2194–2206, 2017.

[24] K. Belharet, D. Folio, and A. Ferreira, "Control of a Magnetic Microrobot Navigating in Microfluidic Arterial Bifurcations through Pulsatile and Viscous Flow," *2012 IEEE/RSJ Int. Conf. Intell. Robot. Syst.*, vol. 1, pp. 2559–2564, 2012.

[25] D. R. Frutiger, K. Vollmers, B. E. Kratochvil, and B. J. Nelson, "Small , Fast , and Under Control : Wireless Resonant Magnetic Micro-agents," *Int. J. Robot. Res. Vol.*, 2010.

[26] P. Vartholomeos, M. R. Akhavan-sharif, and P. E. Dupont, "Motion planning for multiple millimeter-scale magnetic capsules in a fluid environment," *2012 IEEE Int. Conf. Robot.*

Autom., pp. 1927–1932, 2012.

[27] S. Floyd, E. Diller, C. Pawashe, and M. Sitti, "Control methodologies for a heterogeneous group of untethered magnetic micro-robots," *Int. J. Rob. Res.*, 2011.

[28] M. Salehizadeh and E. Diller, "Two-agent formation control of magnetic microrobots in two dimensions," *J. Micro-Bio Robot.*, pp. 9–19, 2017.

[29] R. Pieters, H. Tung, S. Charreyron, D. F. Sargent, and B. J. Nelson, "RodBot : a Rolling Microrobot for Micromanipulation," *2015 IEEE Int. Conf. Robot. Autom.*, pp. 4042–4047, 2015.

[30] L. Arcese, M. Fruchard, A. Ferreira, L. Arcese, M. Fruchard, A. Ferreira, A. Controller, M. Microrobot, L. Arcese, M. Fruchard, and A. Ferreira, "Adaptive Controller and Observer for a Magnetic Microrobot," *EEE Trans. Robot.*, 2013.

[31] I. S. M. Khalil, L. Abelmann, and S. Misra, "Magnetic-Based Motion Control of Paramagnetic Microparticles With Disturbance Compensation," *IEEE Trans. Magn.*, vol. 50, no. 10, pp. 1–10, 2014.

[32] W. Ma, F. Niu, X. Li, H. Ji, J. Yang, and D. Sun, "Modeling and Closed-Loop Control of Electromagnetic Manipulation of A Microparticle," *2015 IEEE Int. Conf. Robot. Autom.*, pp. 143–148, 2015.

[33] J. Wang, N. Jiao, S. Tung, and L. Liu, "Automatic Path Tracking and Target Manipulation of a Magnetic Microrobot," *Micromachines*, pp. 1–14, 2016.

[34] H. Abass, M. Shoukry, and A. Klingner, "Disturbance Observer-Based Motion Control of Paramagnetic Microparticles Against Time-Varying Flow Rates," *2016 6th IEEE Int. Conf. Biomed. Robot. Biomechatronics*, pp. 67–72, 2016.

[35] R. Pieters, S. Lombriser, A. Alvarez-aguirre, and B. J. Nelson, "Model Predictive Control of a Magnetically Guided Rolling Microrobot," *IEEE Robot. Autom. Lett.*, vol. 1, no. 1, pp. 455–460, 2016.

[36] D. Folio, K. Belharet, and A. Ferreira, "Motion control analysis of two magnetic microrobots using the combination of magnetic gradient and oscillatory magnetic field," *2017 Int. Conf. Manip. Autom. Robot. Small Scales*, 2017.

[37] W. Ma, J. Li, S. Member, F. Niu, H. Ji, and D. Sun, "Robust Control to Manipulate a Microparticle with Electromagnetic Coil System," *IEEE Trans. Ind. Electron.*, vol. 64, no. 11, pp. 8566–8577, 2017.

[38] S. Afshar, M. B. Khamesee, and A. Khajepour, "Optimal Configuration for Electromagnets and Coils in Magnetic Actuators," *IEEE Trans. Magn.*, vol. 49, no. 4, pp. 1372–1381, 2013.

[39] A. J. Petruska, J. B. Brink, and J. J. Abbott, "First Demonstration of a Modular and Reconfigurable Magnetic-Manipulation System," *2015 IEEE Int. Conf. Robot. Autom.*, pp. 149–155, 2015.

[40] D. Folio, K. Belharet, and A. Ferreira, "Optimal Control of Multiple Magnetic Microbeads Navigating in Microfluidic Channels," *2016 IEEE Int. Conf. Robot. Autom.*, no. 1, 2016.

[41] S. Floyd, C. Pawashe, and M. Sitti, "Two-Dimensional Contact and Noncontact Micromanipulation in Liquid Using an Untethered Mobile Magnetic Microrobot," *IEEE Trans. Robot.*, vol. 25, no. 6, pp. 1332–1342, 2009.

[42] C. Pawashe, F. Steven, E. Diller, and M. Sitti, "Two-Dimensional Autonomous Microparticle Manipulation Strategies for Magnetic Microrobots in Fluidic Environments," *IEEE Trans. Robot.*, vol. 28, no. 2, pp. 467–477, 2012.

[43] U. Lehmann, S. Hadjidj, V. K. Parashar, A. Rida, and M. A. M. Gijs, "TWO DIMENSIONAL MAGNETIC MANIPULATION OF MICRODROPLETS ON A CHIP," *13th Intemational Conf. Solid-State Sensors, Actuators Microsyst.*, pp. 77–80, 2005.

[44] E. Diller, S. Floyd, C. Pawashe, and M. Sitti, "Control of Multiple Heterogeneous Magnetic Microrobots in Two Dimensions on Nonspecialized Surfaces," *IEEE Trans. Robot.*, vol. 28, no. 1, pp. 172–182, 2012.

[45] K. Belharet, D. Folio, and A. Ferreira, "Three-Dimensional Controlled Motion of a Microrobot using Magnetic Gradients," *Adv. Robot.*, vol. 1864, 2011.

[46] I. S. M. Khalil, R. M. P. Metz, B. A. Reefman, and S. Misra, "Magnetic-Based Minimum Input Motion Control of Paramagnetic Microparticles in Three-Dimensional Space," *2013 IEEE/RSJ*

Int. Conf. Intell. Robot. Syst., pp. 2053–2058, 2013.

[47] S. Schuerle, S. Erni, M. Flink, B. E. Kratochvil, and B. J. Nelson, "Three-Dimensional Magnetic Manipulation of Micro- and Nanostructures for Applications in Life Sciences," *IEEE Trans. Magn.*, vol. 49, no. 2, pp. 321–330, 2013.

[48] E. Diller and M. Sitti, "Three-Dimensional Programmable Assembly by Untethered Magnetic Robotic Micro-Grippers," *Adv. Funct. Mater.*, pp. 4397–4404, 2014.

[49] Z. Zhang, F. Long, and C. Menq, "Three-Dimensional Visual Servo Control of a Magnetically Propelled Microscopic Bead," *IEEE Trans. Robot.*, vol. 29, no. 2, pp. 373–382, 2013.

[50] S. J. Kim, S. M. Jeon, J. K. Nam, and G. H. Jang, "Closed-Loop Control of a Self-Positioning and Rolling Magnetic Microrobot on 3D Thin Surfaces Using Biplane Imaging," *IEEE Trans. Magn.*, vol. 50, no. 11, pp. 1–4, 2014.

[51] E. Diller, J. Giltinan, and M. Sitti, "Independent control of multiple magnetic microrobots in three dimensions," *Int. J. Rob. Res.*, 2013.

[52] I. S. M. Khalil, V. Magdanz, S. Sanchez, and O. G. Schmidt, "The Control of Self-Propelled Microjets Inside a Microchannel With Time-Varying Flow Rates," *IEEE Trans. Robot.*, vol. 30, no. 1, pp. 49–58, 2014.

[53] L. Yang, Q. Wang, and L. Zhang, "Model-Free Trajectory Tracking Control of Two-Particle Magnetic Microrobot," *IEEE Trans. Nanotechnol.*, vol. 17, no. 4, pp. 697–700, 2018.

[54] S. Verma, W. Kim, S. Member, and J. Gu, "Six-Axis Nanopositioning Device With Precision Magnetic Levitation Technology," *IEEE/ASME Trans. Mechatronics*, vol. 9, no. 2, pp. 384–391, 2004.

[55] M. P. Kummer, S. Member, J. J. Abbott, B. E. Kratochvil, R. Borer, A. Sengul, S. Member, B. J. Nelson, and S. Member, "OctoMag : An Electromagnetic System for 5-DOF Wireless Micromanipulation," *IEEE Trans. Robot.*, vol. 26, no. 6, pp. 1006–1017, 2010.

[56] B. Kratochvil, M. Kumar, S. Erni, R. Borer, D. Frutiger, S. Schurle, and B. Nelson, *MiniMag: A Hemispherical Electromagnetic System for 5-DOF WirelessMicromanipulation*. 2014.

[57] E. Diller, J. Giltinan, G. Z. Lum, Z. Ye, and M. Sitti, "Six-Degrees-of-Freedom Remote Actuation of Magnetic Microrobots," *Robot. Sci. Syst. 2014*, 2014.

[58] A. W. Mahoney and J. J. Abbott, "5-DOF Manipulation of an Untethered Magnetic Device in Fluid using a Single Permanent Magnet," *Robot. Sci. Syst. 2014*, 2014.

[59] J. Giltinan, "Simultaneous Six-Degree-of-Freedom Control of a Single-Body Magnetic Microrobot," *IEEE Robot. Autom. Lett.*, vol. 4, no. 2, pp. 508–514, 2019.

[60] E. B. Steager, M. S. Sakar, C. Magee, M. Kennedy, A. Cowley, and V. Kumar, "Automated biomanipulation of single cells using magnetic microrobots," *Int. J. Rob. Res.*, pp. 346–359, 2013.

[61] A. De Langlade, "Position Control of a Magnetic Levitation Device Using a Non-Linear Disturbance Observer and Influence of the Position Sensing," *IECON 2017 - 43rd Annu. Conf. IEEE Ind. Electron. Soc.*, 2017.

Acknowledgment

The authors would like to acknowledge the support provided by Prince Mohammad bin Fahd University.

Grasp Pose Detection Based On Point Cloud Shape Simplification

H W Liu[1] and C Q Cao[2,3*]

[1]School of Mechanical Engineering, Nanjing University of Science & Technology, Nanjing 210094, China;
[2]Anhui Polytechnic University, Wuhu 241000, China;
[3]HIT Wuhu Robot Technology Research Institute Co., LTD, Wuhu 241007, China;

*caochuqing@hitrobot.com.cn

Abstract. In the robot grasping environment, grasping unknown objects that have neither model data nor RGB data is very important, we present a new approach for grasping unknown objects. Firstly, depth camera is used to obtain the partial 3D data of target object. Then, we perform 3D segmentation and the segmented parts are simplified to a cylinder, a sphere, an ellipsoid or a parallelepiped based on the geometric and semantic shape characteristic. Grasping constraints are used to obtain grasp candidates based on the simplified shape. The grasp descriptions are used to describe each candidates and CNN is used for grasping training. We evaluate our grasp networks both in simulation and robotic experiment, and the experiment result shows that the grasp quality score using simplified constraints is more robust than without simplified constraint when the gripper posture is uncertain.

1. Introduction

Autonomous grasping of robotic is one of the core technologies of intelligent robotic. However, the uncertainty of shape and attitude of target object limits the application of robotic grasping technology. To address this problem, recent research has focused on building a large-scale grasping dataset [1] [2] and using neural networks for robotic grasping training, using images data [3] or point cloud data [4] acquired by sensors as input of grasping network. Most current grasping methods use two-dimension (image) data or 2.5D (depth image) data [5] as the input of grasping training. And very few people take the three-dimensional geometry information [6] of objects into consideration. Intuitively, when the shape and posture of the grasped object are uncertain, a single grasping direction may limit the grasping quality of robotic [7]. The grasping quality is usually related to the interaction between the robot gripper and the surface of target object, the lack of object geometric analysis may have a negative effect on grasping.

In order to address the above problems, we use the 3D data of the target object for training. However, computing a stable grasp is a “big-spending” when a large number of novel objects are added to the grasp database. Therefore, reusing the grasps of similar shapes is promising [8][9], as opposed to computing stable grasps for the novel objects from scratch. Those works take the appropriate parts of a given object for grasping, depending on the similar topologies and geometric shapes.

In this paper, we propose a method to detect grasp pose based on the three-dimensional point cloud simplification. As illustrated in Figure 1, first, the objects are segmented based on geometric and semantic shape characteristics, and then the main part is simplified as a kind of sphere, ellipsoid, cylinder and parallelepiped. The grasping pose is constrained based on the simplified shape feature.

Under the grasp constraint, we built a grasp dataset and trained it in YCB dataset [10] – an online dataset that includes depth scanning data of real objects. In our previous work, we have realized the segmentation and simplification of three-dimensional point cloud. Here, we summarize the contributions of this paper in the field of robotic grasping.

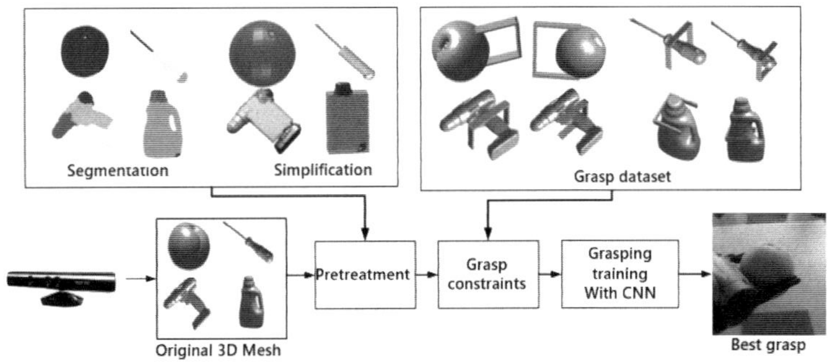

Figure 1. An illustrate of our proposed method for grasping pose detection based on three-dimensional point cloud shape simplification

- We recommend applying the simplified algorithm based on the basic form to describe objects. In order to avoid the singularity of object description and make a more vivid description of object shape characteristics, the sphere, the ellipsoid, the cylinder and the parallelepiped are used to describe objects.
- We propose a grasp strategy to constrain the grasping position based on the simplified shape characteristics and suggest to grasp the main parts of the object. And acquiring grasp candidates based on the grasping constraints.
- We build a new grasp description that incorporates grasp heightmaps and angle graph of object surface to measure the grasp similarity. We build a new grasp description that incorporates grasp heightmaps and angle graph of object surface to measure the grasp similarity.

2. Related work

The autonomous grasping depends on the recognition of object and the description of grasped object shape. The description of the object shape provides an effective way to select grasping strategy. In this section, we give a brief introduce of previous research work on robotic grasping, especially similar objects grasping and part-based grasping.

2.1. Similar objects grasping

In the grasping experiment on similar objects, the grasped objects can be classified into the same categories based on different characteristics, such as usage, application, or geometric shape. Goldfeder et al. [11] proposed a method of using hyperquadric curve tree to automatically approach the object geometry; Lei Q et al. [12] synthesized an executable grasp used cylinder searching on a single point cloud for unknown object. In [13], [14], the authors used the shape similarity of objects to transfer the optimal grasp position to the novel object. Tian H et al. [15] segmented the target objects according to the shape, marked the optimal grasp posture of each sample objects and mapped it to the parts with the same semantics in the target object to obtain the optimal grasp posture.

2.2. Part-based grasping

Many scholars have been proposed to segment and grasp the target parts based on some grasping characteristics. Kai et al. [16] proposed a segmentation algorithm based on the minimum volume bounding box, which decomposition the 3D points into simple objects that composed of multiple bounding boxes; Aleotti et al. [17] proposed a programming-based demonstrated method and a 3D segmentation method, they segmented the grasped object based on Reeb graphs which were used to generate the topological representation. At present, many researches use part information of objects for grasping training. In [18], [19], the authors design a GPD method to normalizes the point cloud and extract the projection features of 3-channels, 12-channels and 15-channels in the grasp closing region,

IOP Conf. Series: Materials Science and Engineering **717** (2020) 012007 doi:10.1088/1757-899X/717/1/012007

and construct a grasping classification model based on CNN. Liang H et al. [20] proposed PointNetGPD grasping method based on PointNet [20], taking the original point cloud as input and directly processing the three-dimension data that locates in the jaw for grasping detection and evaluation.

3. Grasp pose detection

3.1. Definitions

Given a set of point cloud and a description of the grasp configuration, the grasp pose detection problem is to identity grasp configuration form which a stable grasping quality would be formed if the robot's gripping were to close. Given a specified object o, let $\mathbb{Q} \subset \mathbb{R}^3$ denote the robot workspace. The input of the algorithm is a set of 3D point cloud, $C \subset \mathbb{Q}$. Each point in the 3D point data is real-time 3D data acquired by the camera and matching at least one camera viewpoint acquisition data, V: $C \rightarrow v$, where $v \subset \mathbb{R}^3$ indicates the set of camera viewpoints, $\mathrm{CV} = (C, v, V)$ indicates the point cloud after multi-viewpoint data combination and viewpoints respectively. Let $\mathrm{g} = (\mathrm{p}, \mathrm{r}) \subset \mathbb{R}^6$ denote a set of grasp configuration in 3D robot workspace, where $\mathrm{p} = (x, y, z) \subset \mathbb{R}^3$ denote the position of the gripper, $\mathrm{r} = (r_x, r_y, r_z) \subset \mathbb{R}^3$ denote the orientation of the gripper.

The expected grasping result is that if the robotic gripper is moved to a given gripper pose g, when the gripper figure are closed, the force closure of the grasped object can be achieved. In this paper, we only consider two-fingered parallel jaw gripper and assume a depth camera to obtain real-time depth data.

3.2. Outline of the algorithm

In order to achieve the robust grasp and obtain the appropriate grasp pose, firstly, we pretreated the 3D data to achieve segmentation and simplification. Grasp constraints are used to obtain grasp candidates based on the simplified shape. The grasp descriptions in this paper are used to describe each candidates and CNN is used for grasping training. Finally, executing the grasp posture with highest score. The specific steps are shown in Figure 1.

In our previous work [21], we have completed the steps of pretreatment and successfully simplified the segmented parts into one of sphere, ellipsoid, cylinder and parallelepiped. As shown in Figure 2, our algorithm can be implemented in different objects, such as regular objects (plane, bottle *et al*) and irregular objects (cat, chicken model *et al*), the shown model in Figure 2 is acquired in SHREC'15 datasets [22].

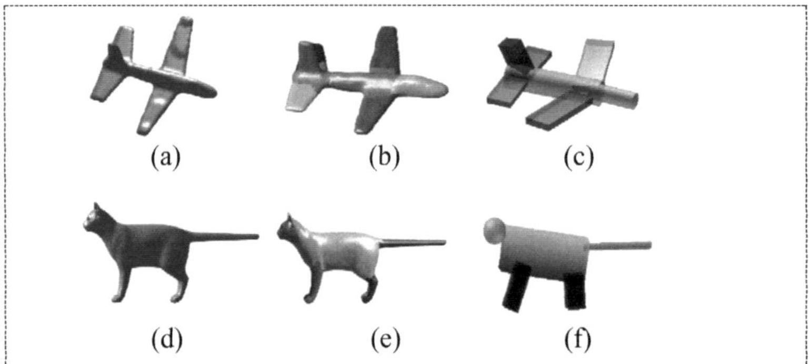

Figure 2. Pretreatment example diagrams. (a)(d) The original data of plane and cat. (b) (e) Segmentation. (c)(f) Simplification

It should be noted that in this grasping algorithm, we only simplify the main part of segmentation parts in order to reduce the running time, and focus on the next steps.

3.3. Sampling grasp candidates

The goal of this section is to obtain a lager set of grasp candidates (6-DOF hand pose), which may contain one or more optimal grasps. When selecting grasp candidates, we hope they can be evenly distributed among the effective parts, that is, the graspable part of object surface. In this experiment, we defined each grasp candidate, $h \in \mathbb{Z}$, as a simple "hand", where $\mathbb{Z}$ is the region of interest.

In the pretreatment step, we have simplified the main part of target object. We set the main part as the region of interest and constrained the grasping pose based on the simplified shape features to obtain the grasp candidates. The following experiment confirmed this method is effective for robot grasping. The grasp constraint is shown below:

- Simplified into sphere: the center position p_{cen} of the gripper is located in the center of the ball, the orientation r_{cen} of the gripper is a random direction vector (shown in Figure 4(a)).
- Simplified into ellipsoid: the center position p_{cen} of the gripper is located in the central long axis of the ellipsoid, the orientation r_{cen} is the same as one of the two short central axis (shown in Figure 4(b)).
- Simplified into cylinder: the center position p_{cen} of the gripper is at the central axis, the orientation r_{cen} is parallel to the central axis (shown in Figure 4(c)).
- Simplified into parallelepiped: the center position p_{cen} of the gripper is at the center long axis, the orientation r_{cen} is the same as one of all edges except the shortest edge of the parallelepiped (shown in Figure 4(d)).

We define $D(h) \in \mathbb{Q}$ as the volume occupied by the robotic gripper when it is fully open. Defined closing region , $C(h) \in \mathbb{Q}$, as the volume area swept by the gripper when the fingers are closed. In the vertical direction of each gripper, we "push" the gripper until the following conditions are met, and Figure 3 shows the above conditions occurred when grasping object.

Condition 1. *The grabs hand and the point cloud will not collide and contact during the process of "push" hand :* $D(h) \bigcap C = \varnothing$, *and the center of the* $C(h)$ *is* p_{cen}.

Condition 2. *The side of the gripper hand or the bottom of the hand collides with the point cloud:* $D(h) \bigcap C \neq \varnothing$, *then stops pushing forward.*

Figure 3. Illustrate of the conditions present in the paper. (a) Condition 1 occurred when grasping. (b) Condition 2 occurred when grasping.

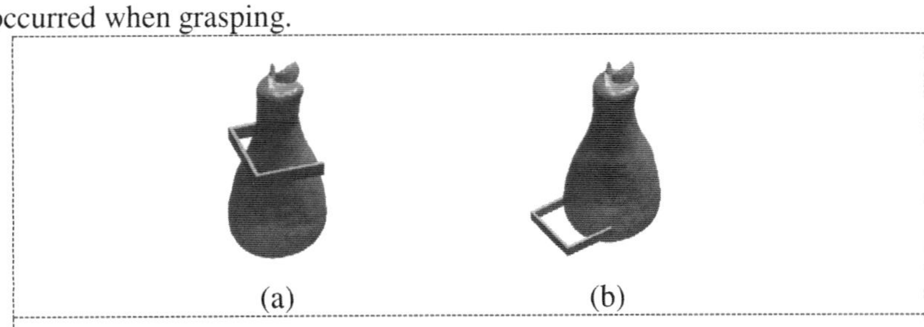

Figure 3. Illustrate of the conditions present in the paper.

(a) Condition 1 occurred when grasping

(b) Condition 2 occurred when grasping

On the basis of grasping constraints, considering the noise of the input cloud and obtaining sufficient grasp candidates, all the gripper positions p_{cen} have random errors of 0~1 cm and all the gripper orientations r_{cen} have random errors of 0~15 degree. As illustrated in Figure 4, we obtain grasping candidates though the above grasp constraints.

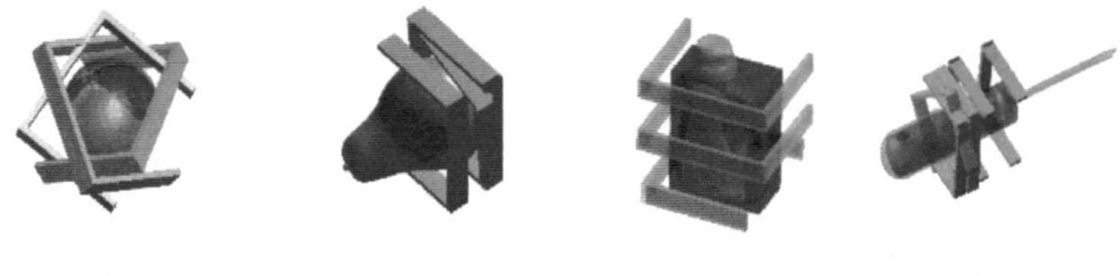

(a) (b) (c) (d)

Figure 4. Illustrate of grasping candidates obtained using our algorithm

3.4. Classifying grasp candidates

The grasp training network in this paper is defined as a binary classification task that the input is the feature representation of a set of grasp candidates and the output is a prediction of whether or not the grasp candidate is graspable. The network structure we use is the same as the LeNet-5 [23] CNN structure: three convolutional layers, two pooling layers and a fully connected layer with a softmax on the output. Except the size of network learning rate is 0.01, the size of output, pooling strides and kernel, *etc.* are all the same as the LeNet network framework provided in Caffe [24].

3.5. Grasp candidate representation

We use the geometric of the mesh surfaces that contained in the closing region $C(h) \subset \mathbb{Q}$ to represent a grasp candidate. The two fingers of the robotic grab can be close to or far away from each other, because they are modeled as rectangles in this experiment, and the shape of the closing region $C(h) \in \mathbb{Q}$ is a cuboid. Figure 5(a) shows a grasp candidate generated with three-dimensional mesh data from YCB dataset [10]. The region shown in red is the mesh points within closing region $C(h)$. We define $C(h) = (\mathrm{c}, \mathrm{v})$ be a closing region, $C(h)$ parameter is composed of the gripper centroid in 3D space $\mathrm{c} \in \mathbb{R}^3$ and the approach direction of the gripper, or direction along which the fingers close, $\mathrm{v} \in \mathbb{C}^2$. First, the mesh points in the closing region $C(h)$ are transformed into the unified grasp coordinate, as introduced in Figure 5(b) where z axis is parallel to v, Figure 5(b) shows the closing region $C(h)$ and local gripper coordinate.

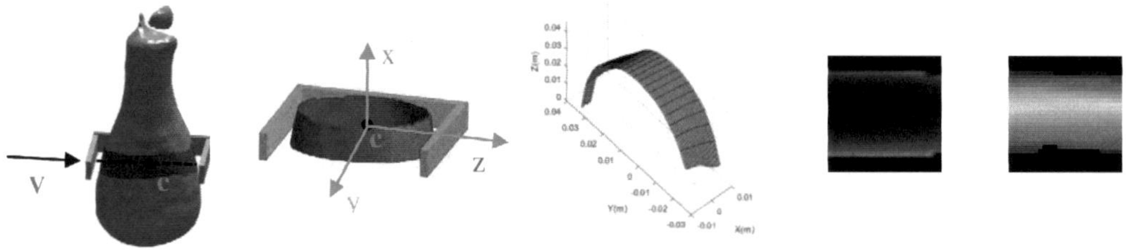

(a) (b) (c) (d) (e)

Figure 5. Grasp representation. (a) A grasp candidate generated from partial grid data. (b) Local voxel grid frame. (c) Minimum projection distance long z axis. (d-e) Example of grasp images used as input to the classifier

Because we evaluate each grasp candidate as a high grasping quality or not by judging whether it is a force closure grasp [25], we only consider the surface that is in contact with two fingers of the jaw. To measure grasp similarly, we embed each grasp candidate $C(h) = (\mathrm{c}, \mathrm{v})$ in a feature space based on 2D projection heightmaps $h(x, y)$ and angle graph $a(x, y)$ of the local surface on left and right fingers of the jaw. Take the right jaw finger as an example, voxel the closing region in the xy plane to 30×30. As shown in Figure 5(c), let $r \subset \mathbb{R}$ be a minimum projection distance long z direction, and the heightmaps $h(x, y)$ is obtained through equation 1. Let the $v_1 \subset \mathbb{C}^2$ be the normal vector of the minimum projection distance point, and calculate the angle $w_1 \subset \mathbb{R}$, $w_1 \in [0, 90]$ between v_1 and the z axis, and obtain the angle graph $a(x, y)$ (equation 2).

$$h(x, y) = \frac{\max(r) - r(x, y)}{\max(r)} \times 255 \tag{1}$$

$$a(x,y)=\frac{90-w_1(x,y)}{90}\times 255 \tag{2}$$

The two images, h and a, are 30×30 images (Figure 5(d,e)). In summary, for each of the projection in two directions, we have two channels of feature representation respectively, and a total of 4 channels.

3.6. Generating training labels

We trained network in the follow form: First, we use the method presented in Section 3.3 to constrain the grasping pose based on the simplified shape to obtain the grasp candidates. Then, we obtain the corresponding surface features of each grasp candidate by using the method described in Section 3.5, and attach a label to each grasp candidate to indicate whether or not the candidate is an effective grasp pose by evaluating a force closure grasp [24] would be obtained if the grippers were to close, that is, each of the surface normals at the contacts between gripper and grasped object is opposite to the finger closing direction v and collinear with the connection line between the contacts (see Figure 6).

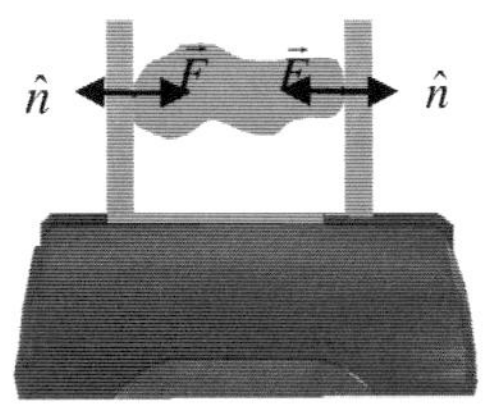

Figure 6. A force closure grasp, with contact surface normals antiparallel to each other.

Unfortunately, it is very tricky to determine whether a grasp candidate is a force closure grasp, because the real-time mesh is noise-containing. For example, the mesh data in the YCB dataset is noise-containing because it is reconstructed from the actual sensor. We "softening" the force closure condition in order to address the above problems. In particular, we consider the point within 2mm of the contact point between two fingers when the fingers are closed after voxelized the point cloud. When the normal of more than 1/4 point that exists in each contact is less than 20 degree with the finger closing direction v, it can be judged as graspable. The above method aims to search for a force closure grasp in small regions that may establish contact.

4. Experiment

We carry out the simulation experiment and the robotic grasping experiment on the method proposed in this paper respectively. In the simulation experiment, we mainly evaluate the performance of pretreatment method and grasp classification network. In the robotic experiment, we conduct multiple robotic grasp experiments to determine whether our grasping model can be implemented in real world.

4.1. Simulation experiments

In the simulation experiment, the grasping dataset we use is based on the YCB dataset, which contains the mesh points of 77 different objects formed by scanning and processing in real-time, and matching the grasp candidate representation obtained by the YCB dataset with the corresponding grasp labels that indicate whether or not the candidate is graspable. We defined robot grip as a parallel jaw that can grasp objects up to 11cm. The compute system we use consists of 16 GB of system memory, Corei7-8700k CPU, and an NVIDIA GeForce RTX 2080 graphic card. We build a dataset with 100k candidates, in which the positive and negative samples are 80k/20k respectively, and it takes 40min to train. In order to verify the effectiveness of the proposed method, we analyze the pretreatment method and grasping training method respectively.

1) Comparison between pretreatment and unpretreatment: The method we proposed in this paper pretreatments the object and then builds a new grasp description. In order to evaluate the positive effects of the pretreatment method proposed in this paper, we performed two sets of candidate experiments respectively using YCB dataset. In addition to our previous 100k grasp candidates, we recreated a dataset of grasp candidates that randomly acquired approximate 2k grasp candidates for each of the 77 YCB objects, with a balance between positives and negatives of the grasp description, we obtained a total of 150k candidates. We obtained approximately 94% accuracy over 77 objects in the YCB dataset when training 100k grasp candidates with pretreatment (the red line in Figure 7). We obtain only 89% accuracy (the blue line in Figure 7) when we train 150k grasp candidates without pretreatment. The accuracy of two grasping methods is demonstrated in Figure 7.

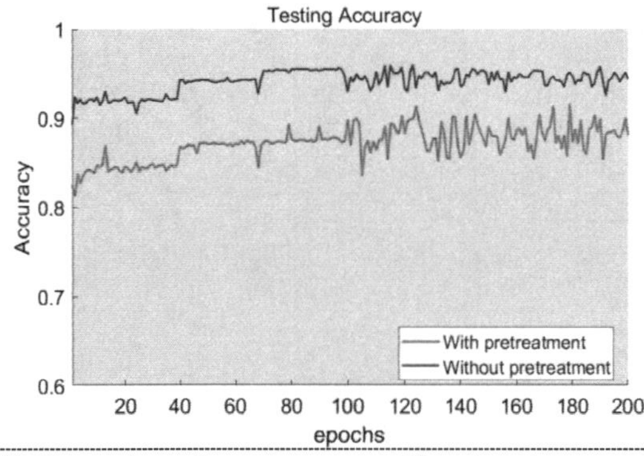

Figure 7. Grasp detection accuracy without pretreatment (blue); grasp detection accuracy with pretreatment (red)

From Figure 7 we can see that the training model with pretreatment obtains better result than without pretreatment. It allows us to use pretreatment to acquire varying degrees of prior knowledge, which has a positive effect on the robotic grasping judgement. In the simulation experiment with pretreatment, since we have obtained grasping candidates which have a higher probability of grasping based on the object shape, we obtained higher grasp accuracy with fewer grasp candidates. Depending on the results described above, pretreatment plays a positive role in robotic grasping judgement.

2) Compared with other grasp methods: In the comparative experiment, we mainly want to compare the advanced grasping methods with our proposed methods in terms of performances on grasping quality classification. We chose GPD [18] [19] and PointNetGPD [20] as baseline because they are all the works closest to ours and using 3D point clouds as input and training. In the experiment, since we cannot obtain the three-dimension data of the target object's unobserved area from the camera location to calculate the 15-channel version of GPD, we only compare the 3-channel, 12-channel of GPD methods and PointNetGPD method. We train at different training set size for the same dataset (YCB dataset), and list the best grasp accuracy result and average grasp accuracy result among the 200 epochs in Table 1. We compare these representations by evaluating training set size and accuracy with which they can predict grasps.

Table 1. Accuracy of different model and configurations

Training model	**Training set size**	**Grasp accuracy**	
		Best result	Average result
Our method	0.1M	95.74%	94.06%
PointNetGPD	1.6M	92.18%	91.81%
GPD (12channel)	3.64M	84.29%	83.50%
GPD (3-channel)	3.63M	82.50%	81.38%

From the above experimental results, we can draw the following conclusions. First, under the same dataset, our and PointNetGPD methods perform significantly better than all the GPD methods in grasp accuracy. Both of them have more than 90% grasp accuracy, and require less data for training than GPD. Furthermore, although we have similar results with the grasp accuracy of PointNetGPD, the performance of our method is better than PointNetGPD method in some fields, we need fewer training set size compared with PointNetGPD method in training set size. And the method we proposed has an average 3% improvement over the PointNetGPD baseline. As Table 1 illustrated that our proposed method manifests a higher grasp accuracy in best result and average result, which indicates that our proposed grasping model can obtain more accurate graspable region of the target object based on the geometric characteristics of local point clouds.

4.2. Robotic experiment

To validate our algorithm, we use the Baxter robotic for grasping in the experiment. The robotic has 7-DOF arm and is equipped with electric parallel jaws with a width limit of 3 to 11cm. Each gripper is connected with two force sensors to measure the force between the two gripper arms. In order to obtain the 3D data of grasped object, we mount Kinect v1 sensor to the robot's left wrist, and the robot's right jaw for grasping (shown in Figure 8(a)). The robot communications with the PC through the robot operating system (ROS), and collision-free inverse kinematics solver IK [24] is used for solving inverse kinematics.

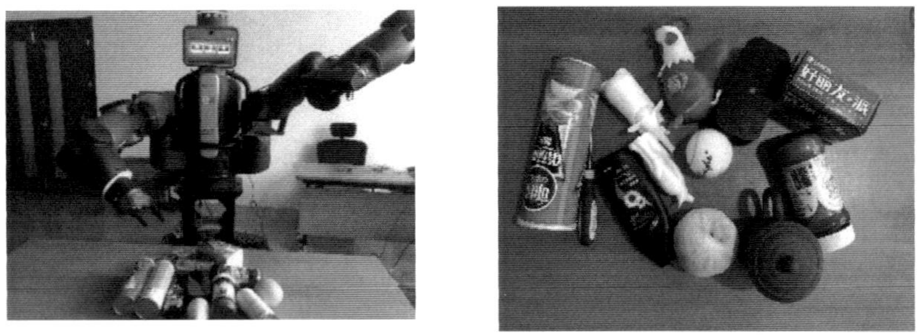

(a) (b)

Figure 8. Configuration of robotic grasping experiments. (a) Grasping experiment device with a Baxter robot and mount Kinect v1 sensor to the robot's left wrist. (b) The target objects in our robotic experiments.

We select 12 common household items, and 6 of which are contained in the grasp dataset, such as peach, chips can *et al*, while the rest are novel objects, such as shampoo, chess model *et al*. The objects are shown in Figure 8(b). Each object for grasping is placed on the desktop separately.

In order to get as much multi-angle point cloud data as possible, when using depth camera to collect point cloud, the robotic arm always performs a collision-free trajectory movement [26] on a hemisphere centered on the target object. The radius of the hemisphere is 80cm and the arm rotates 100 degrees around the hemisphere center. At all times, the lens of depth sensor always points towards the center point of desktop workspace and the angle to the desktop is 50 degrees. We obtain real-time 3D reconstruction data of grasped scene through KinectFusion [27]. And we pretreatment the 3D points cloud of target object (shown in Figure 9(b)) by fitting a plane to the table using RANSAC [28] and extracting the point (shown in Figure 9(c)) above the table based on Euclidean distance clustering. Then, we use the pretreatment method and the grasping training neural network mentioned above to generate best grasps for testing. According to shape constraints, candidate with the highest grasping quality is selected as the best grasp (shown in Figure 9(e)). An example of grasping chess model is demonstrated in Figure 9.

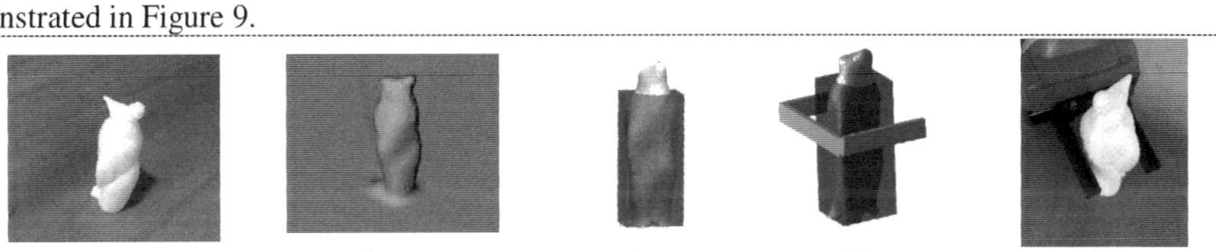

(a) (b) (c) (d) (e)

Figure 9. An example of grasping chess model. (a) Model on the table. (b) Real-time acquisition of muti-angle 3D reconstruction data through KinectFusion. (c) Extraction of model 3D point data. (d) Best grasp candidate after shape constraint. (e) Baxter robotic grasps model

In this experimental, we ran 120 clutter experiments. We performed grasp tests on the object shown in Figure 8(b). Each object is grasped 10 times, and the position and posture of object in each grasp are random and different. In the grasping experiment, we defined a "successful raise" as the jaw grasped and raised the object 20cm without dropping it. And a grasp is successful if (1) the gripper successfully grasps the object and (2) it is a "successful raise". Table 2 demonstrates the accuracy of grasping dataset objects and novel objects respectively.

Table 2. Accuracy of grasping dataset and novel objects

Object type	Accuracy
Dataset objects	93.3%
Novel objects	91.6%
Total	92.5%

Among the 60 grasping of dataset objects, high grasping success accuracy 100% has been generated because the objects are regular, such as chips can and pudding box. There were several unsuccessful grasps when grasping peach and screwdriver, the narrow width of claw and low height of the acquired point cloud probably lead to several failed grasps. In the experiment of grasping novel objects, there were several failed grasps when grasping chicken model and elephant model, the poor quality and the irregular shape may lead to several failed grasps. It can be concluded that our algorithm can get robust when grasping regular shape objects. For irregular objects, we may get wrong grasping result, but the failed rate is limited. Overall the grasping success accuracy of robotic remains above 90%, which indicates that our proposed grasp model can effectively grasp objects.

5. Conclusion

In this paper, we train robotic grasping based on previous pretreatment work that first segmentation and simplification point cloud, and a large set of grasp candidates are obtained by grasping constraint based on pretreatment result and then the grasping classification is trained by neural network. We proposed three ways to increase the robotic understanding of target object. First, in order to avoid the singularity of object description and to make a more vivid description of object shape characteristics, we proposed a pretreatment method to apply the simplification of the basic form to the description of objects. Second, we propose a grasp strategy based on the basic shape and suggest to grasp the main parts of the object. Third, we build a new grasp description to measure the grasp similarity. Compared with current grasp quality classification methods, our experiments in this paper show an increase of about 3% in grasp accuracy and reach 94% grasp accuracy. This result emphasizes that the way we proposed to enhance the classification accuracy and robotic grasping accuracy.

6. References

[1] Mahler J, Pokorny F T and Hou B 2016 Dex-Net 1.0: A cloud-based network of 3D objects for robust grasp planning using a Multi-Armed Bandit model with correlated rewards *IEEE Int. Conf. on Robotics & Automation*

[2] Mahler J, Liang J and Niyaz S 2017 Dex-Net 2.0: Deep Learning to Plan Robust Grasps with Synthetic Point Clouds and Analytic Grasp Metrics *arXiv preprint*

[3] Lenz I, Lee H and Saxena A 2015 Deep learning for detecting robotic grasps *The Int. J. of Robotics Research* **34(4-5)** pp 705-724.

[4] Ten Pas A and Platt R 2015 Using Geometry to Detect Grasps in 3D Point Clouds *Computer Science*

[5] Johns E, Leutenegger S and Davison A J 2016 Deep Learning a Grasp Function for Grasping under Gripper Pose Uncertainty *IEEE/RSJ Int. Conf. on Intelligent Robots and Systems (IROS)* pp 4461-68

[6] Yan X, Hsu J and Khansari M 2018 Learning 6-DOF Grasping Interaction via Deep Geometry-aware 3D Representations *IEEE Int. Conf. on Robotics and Automation (ICRA)* pp 1-9

[7] Ten Pas A, Gualtieri M and Saenko K 2017 Grasp Pose Detection in Point Clouds[J]. The International Journal of Robotics Research *The Int. J. of Robotics Research* **36(13-14)** pp 1455-73

[8] Vahrenkamp N, Westkamp L and Yamanobe N 2016 Part-based grasp planning for familiar objects *IEEE-RAS Int. Conf. on Humanoid Robots* pp 919-925

[9] Stouraitis T, Hillenbrand U and Roa M A 2015 Functional power grasps transferred through warping and replanning *IEEE Int. Conf. on Robotics & Automation* pp 4933-40
[10] Berk Calli, Arjun Singh, James Bruce, Aaron Walsman, Kurt Konolige, Siddhartha Srinivasa, Pieter Abbeel and Aaron M D 2017 Yale-CMU-Berkeley dataset for robotic manipulation research *The Int. J. of Robotics Research* **36** pp 261-268
[11] Goldfeder, Allen and Lackner 2007 Grasp Planning via Decomposition Trees *IEEE Int. Conf. on Robotics & Automation* pp 4679-84
[12] Lei Q and Wisse M 2017 Fast grasping of unknown objects using cylinder searching on a single point cloud *9th Int. Conf. on Machine Vision* **10341** p 1034108
[13] C. Goldfeder, M. Ciocarlie, H. Dang, and P. K. Allen 2009 The columbia grasp database *IEEE Int. Conf. on Robotics and Automation* pp 1710-16
[14] Y. Li, J. L. Fu, and N. S. Pollard 2007 Data-driven grasp synthesis using shape matching and task-based pruning *IEEE Trans. on Visualization and Computer Graphics* **13** pp 732–747
[15] Tian H, Wang C and Manocha D 2019 Transferring Grasp Configurations using Active Learning and Local Replanning *Int. Conf. on Robotics and Automation* pp 1622-28
[16] Kai H, Ruthotto S and Kragic D 2008 Minimum volume bounding box decomposition for shape approximation in robot grasping *IEEE Int. Conf. on Robotics & Automation* pp 1628-33
[17] Aleotti and S. Caselli 2011 Part-based robot grasp planning from human demonstration *IEEE Int. Conf. on Robotics and Automation* pp 4554–4560
[18] Ten Pas A, Gualtieri M and Saenko K 2017 Grasp Pose Detection in Point Clouds *The Int. J. of Robotics Research* **36** pp 1455-73
[19] Gualtieri M, Ten Pas A and Saenko K 2016 High precision grasp pose detection in dense clutter *IEEE/RSJ Int. Conf. on Intelligent Robots and Systems* pp 598-605
[20] Liang H, Ma X and Li S 2018 PointNetGPD: Detecting Grasp Configurations from Point Sets *Int. Conf. on Robotics and Automation* pp 3629-35
[21] Liu H W, Cao C Q and Wang Y J 2019 Autonomous grasping method based on non-structural basic composition analysis *Robot* **41(05)** pp 583-590
[22] Pickup D, Sun X and Rosin P L 2015 Canonical Forms for Non-Rigid 3D Shape Retrieval *Eurographics Workshop on 3d Object Retrieval*
[23] Haykin S and Kosko B 2009 Gradient Based Learning Applied to Document Recognition *Wiley-IEEE Press*
[24] Murray R M, Sastry S S and Zexiang L 2017 A Mathematical Introduction to Robotic Manipulation *CRC Press*
[25] Schulman J, Ho J and Lee A X 2013 Finding Locally Optimal, Collision-Free Trajectories with Sequential Convex Optimization *Robotics: science and systems* **9(1)** pp 1-10.
[26] S. Starke, N. Hendrich, D. Krupke, and J. Zhang 2017 Evolutionary Multi-Objective Inverse Kinematics on Highly Articulated and Humanoid Robots *IEEE Int. Conf. on Intelligent Robots and Systems* pp 6959-66
[27] Newcombe R A, Izadi S and Hilliges O 2012 KinectFusion: Real-Time Dense Surface Mapping and Tracking *Ismar Basel Switzerland Oct* **4(6)** pp 127-136.
[28] Li L, Yang F and Zhu H 2017 An Improved RANSAC for 3D Point Cloud Plane Segmentation Based on Normal Distribution Transformation Cells *Remote Sensing* **9(5)** p 433

Acknowledgments
Our research has been supported in part by National Natural Science Foundation of China (No.61503186).

IOP Conf. Series: Materials Science and Engineering **717** (2020) 012008 doi:10.1088/1757-899X/717/1/012008

An Automatic Calibration Method for Puncture Surgical Robot

H X Yuan, Q Q Li, H Zhao, L J Wang and R Song

School of Control Science and Engineering, Shandong University, Jinan 250061, China

rsong@sdu.edu.cn

Abstract. This paper proposes a rapid and automatic calibration method for puncture surgery robot, via which the hand-eye and tool calibration can be accomplished automatically in a short time. A new method for coordinate conversion between robot flange and needle tip is proposed, based on the method of spherical fitting and singular value decomposition. The experimental results show that the calibration accuracy and stability of this method can meet clinical requirements. There are good prospects for this method using in surgical calibration application.

1. Introduction

With the development of computer aided surgery (CAS) and robotic technologies, surgical robots are increasingly used in the clinical field, and show great advantages in terms of flexibility, stability and accuracy in the field of surgery [1-2]. The puncture procedure is a common and important operation in minimally invasive surgery, where the robot performed can improve the operating precision [3-4], reducing operator visual error and the elimination of the risk of radiation doctors.

At present, the image-guided surgical robot system has been widely used. It uses a six-degree-of-freedom lightweight robot arm as the actuator, the infrared stereo camera as the vision sensor for real-time intraoperative tracking and feedback, and the end flange is equipped with different surgical instruments as the tools of the robot system [5]. There are many researches on surgical robot system. Zeng [6] proposed a calibration method for the hand-eye of surgical robots, which can eliminate the data with large measurement error and improve the calibration accuracy, however it cannot avoid the human operation in the calibration process, and introduce a certain amount of human error. Liu [7] mentioned in the next phase of work to design an automatic calibration system. The latest research results of Niu [8] also emphasize the rapidity of self-calibration. Obviously, improving the automation of calibration is the main trend of current research. However, hand-eye calibration and tool calibration are usually carried out separately, and mass of data is needed to ensure the accuracy, which

IOP Conf. Series: Materials Science and Engineering **717** (2020) 012008 doi:10.1088/1757-899X/717/1/012008

will take up a lot of time.

In order to make up for the complicated operation, long duration, and human intervention in the calibration process, this paper designs a set of automatic calibration system for puncture assisted surgical robot, which realizes the automation of hand-eye calibration and tool calibration of surgical robot system guided by infrared camera. It can not only ensure high accuracy of the system, but also improve calibration efficiency. The experimental results show that the repositioning precision of our method is 0.91 ± 0.03 mm, and the whole calibration process takes about one minute, so that, with the help of the new calibration system, the surgery robot can be better applied in clinic.

2. Calibration approach

2.1. System Establishment

The puncture surgical robot system usually consists of four parts: a workstation for multi-source data processing and robot control, a navigation system used for locating and tracking, an actuator served by a 6-DOF robot arm, and a customized needle equipped with infrared marker balls. In this paper, the mechanical arm uses UR5 (Universal Robots, Denmark), navigation camera uses NDI Polaris (Canada). The robotic surgical puncture system is shown in figure 1.

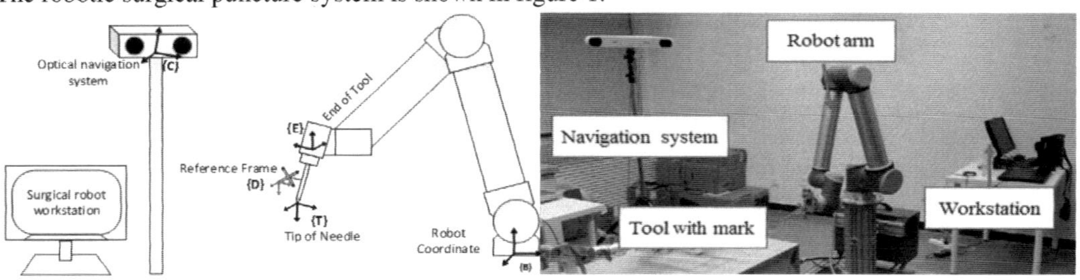

Figure 1. Space axis of the robot system

The role of the calibration can accomplish the unification of four parts coordinate system of the puncture surgical robot, especially the coordinate unification of the navigation camera, mechanical arm and the puncture needle. And this process is generally completed by two calibration methods, which the coordinate unification between the navigation system and the mechanical arm is defined as hand-eye calibration and the coordinate unification between the mechanical arm and the puncture needle is defined as tool calibration. In this paper, the main contribution is to realize the two calibration processes automatically through an operation process. It mainly includes locating the needle tip by the marker balls and using SVD (singular value decomposition) to solve the coordinate transformation.

2.2. Needle tip localization

The coordinate origin of reference frame {D} is obtained in real-time by four infrared reflective balls with relatively fixed positions in the optical camera coordinate system. Supposing that one of infrared reflective ball is the origin of coordinate, the camera can obtain the position and posture information of the reference frame through the information of these infrared balls.

The coordinate of Tip of Needle {T} is established with the probe tip as the coordinate origin, which the direction of the tip is determined by the reference frame. We use the spherical fitting method to locate the needle tip. The piercing tool rotates around its point, recording the position information of infrared ball on the puncture probe. Because of the unique positional relation between infrared ball and needle tip, these recording spots form a sphere. The core of this sphere is the end of the tool center, and its radius is the distance between the origins of the coordinates. Supposing that the ball's center in the optical locator coordinate system is (a, b, c) and its radius is R, the parametric equation for the sphere is:

$$R^2 = (x-a)^2 + (y-b)^2 + (z-c)^2 \tag{1}$$

In order to get the best fitting line experimentally, Equation (2) must be satisfied:

$$minf(x,y,z) = (x-a)^2 + (y-b)^2 + (z-c)^2 - R^2 \tag{2}$$

By this method, the position and posture of the puncture tool tip in the optical camera system can be acquired. Changing the pose from Euler angle to the rotation matrix, we can obtain the transfer matrix ${}^{C}_{T}\mathrm{T}$ between tip of needle {T} and optical camera coordinate {C}.

2.3. Calibration principle based on SVD

The automatic calibration process consists of hand-eye calibration and tool calibration. The hand-eye calibration needs to calculate the transfer matrix ${}^{C}_{B}T$ between the optical camera coordinate system {C} and the robot coordinate system {B}, and the tool calibration needs to calculate the transfer matrix ${}^{E}_{T}T$ between the TCP (Tool Center Point) coordinate system {E} and the puncture tool coordinate system {T}. To establish the transfer matrix between the camera coordinate system and the robot coordinate system, it is necessary to control the movement of the robot arm at least four different positions, recording the center point of the end of the arm $\{{}^{C}_{E}\mathrm{P}\}$ and the end point of the arm $\{{}^{B}_{E}P\}$ directly acquired by the NDI camera at different positions. Set these two sets of three-dimensional point as $\{P_i\}$ and $\{P_i'\}$. ARUN [9] present an algorithm for finding the least-squares solution of R and T, which is based on the SVD of a 3×3 matrix. By this method, it's easy to find the transfer matrix ${}^{C}_{B}T$.

Based on analysis positive and reversal motion of UR robot, we can derive the transfer matrix ${}^{B}_{E}T$. Then we can get the transfer matrix ${}^{B}_{T}T$ and ${}^{E}_{T}T$ by equation (3) (4).

$${}^{C}_{B}T^{-1} \times {}^{C}_{T}T = {}^{B}_{C}T \times {}^{C}_{T}T = {}^{B}_{T}T \tag{3}$$

$${}^{B}_{E}T^{-1} \times {}^{B}_{T}T = {}^{E}_{B}T \times {}^{B}_{T}T = {}^{E}_{T}T \tag{4}$$

2.4. The optimized process

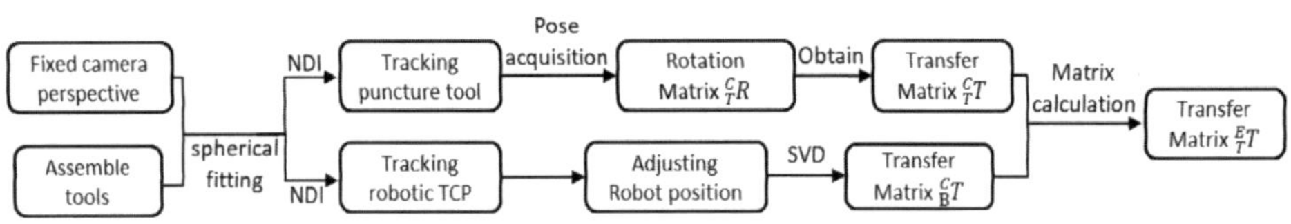

Figure 2. Calibration process

Calibration process steps as shown in figure 2. Before the beginning of automatic calibration, optical camera obtains the coordinates of infrared balls in real time, and the puncture tool rotates around its needle tip while recording the position of infrared sphere. Combined with the spherical fitting method to calculate the position and posture of the needle tip in the camera coordinate system. The hand-eye calibration process needs to control the robot arm to move three times in the camera space. When the robot moves to different locations, there will be a short pause. And we need to record the data from the robot and the optical camera about TCP location. Based on these data, we can obtain the hand-eye calibration transfer matrix by the method of SVD. Also, tool calibration needs to obtain the posture of the needle tip in the camera coordinate firstly. Then, converting the location information to a matrix and the tool calibration transfer matrix is derived by using the result of hand-eye calibration. Compared with the traditional mechanical calibration process, the system can be better integrated with the real scene. The tracking of the puncture tool and the TCP location of robot are completed before calibration which could effectively reduce the operation time.

3. Experiments and results

3.1. Accuracy verification experiment

Accuracy experiment is to study the error change in the process of robot moving. The auxiliary probe is used to randomly take 25 consecutive points in the camera space as the moving target points of the puncture tool, and the robot arm is controlled to move to the points. According to hand-eye calibration matrix ${}^{C}_{B}T$ and tool calibration matrix ${}^{E}_{T}T$, the position of TCP in robot coordinate ${}^{B}_{E}P$ can be converted into the position of needle tip in camera coordinate ${}^{C}_{T}P$ by equation (5).

$$ {}^{C}_{T}P = {}^{C}_{B}T * {}^{B}_{E}P * {}^{E}_{T}T \quad (5) $$

Compare the calculation results with the information tracked by the camera and the values recorded by the auxiliary probe, so the accuracy of the calibration method can be verified. The results are as follows:

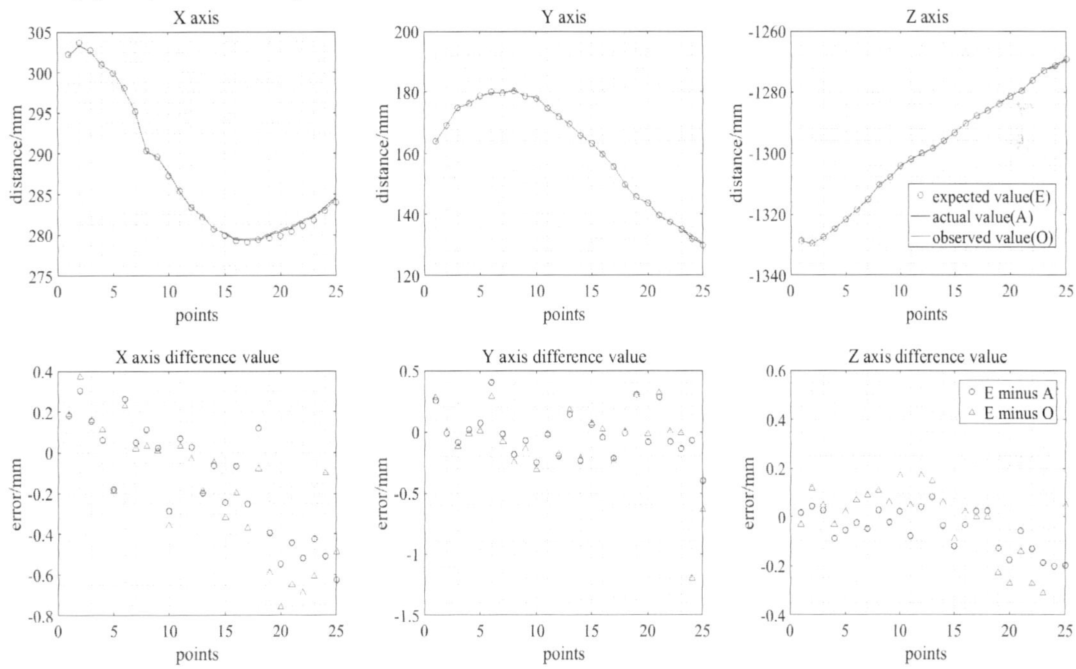

Figure 3. Accuracy verification experiment result

In figure 3, expected value (E) is acquired by auxiliary probe in camera coordinate system, actual value (A) is the result calculated by equation (5), observed value (O) is the position of needle tip obtained by the camera. By analyzing the data, we can see the error range is around 0.1-0.3mm. And the expected value and the actual value are more similar and more stable.

3.2. Stability verification experiment

Stability experiments are designed to verify how the errors of the system will change after the repeated experiments. Coordinate (209.12, -108.63, -1413.97) is set as the target point. Then the robot repeats the automatic calibration process 10 times with different initial postures and control the puncture needle to reach the target point. Experimental data are as follows, where the error result is obtained by subtracting the target position from the actual position. Each actual point is represented as (x, y, z), the error is expressed as ($\boldsymbol{\delta_x}$, $\boldsymbol{\delta_y}$, $\boldsymbol{\delta_z}$) and Err3d is the square root of the error.

Table 1. Stability verification experiment data

	(x, y, z)	($\boldsymbol{\delta_x}$, $\boldsymbol{\delta_y}$, $\boldsymbol{\delta_z}$)	Err3d
1	(209.89, -108.63, -1413.22)	(0.77, 0, 0.75)	1.074
2	(209.48, -108.52, -1413.20)	(0.36, 0.11, 0.77)	0.857
3	(208.99, -109.18, -1413.93)	(0.87, -0.55, 0.04)	1.030
4	(209.34, -109.06, -1413.27)	(0.22, -0.43, 0.70)	0.850
5	(209.14, -108.98, -1413.39)	(0.02, -0.35, 0.58)	0.677
6	(209.70, -108.84, -1413.15)	(0.68, -0.21, 0.82)	1.085
7	(209.63, -108.81, -1412.88)	(0.51, -0.18, 1.09)	1.216
8	(209.26, -107.94, -1413.76)	(0.14, 0.69, 0.23)	0.740
9	(209.32, -108.58, -1413.06)	(0.20, 0.05, 0.91)	0.933
10	(209.03, -108.93, -1413.35)	(-0.08, -0.30, 0.62)	0.693

It can be seen from the data in the table that the error between the actual coordinate point and the target coordinate point is within 1 mm, but the value of the 10 sets of results varies greatly. The main reasons for the large fluctuation of the value is that we can't ensure that the puncture probe revolves accurately around the needle tip when using the sphere fitting method.

4. Conclusions

Considering the problems of complicated operation, long duration and human intervention, an automatic calibration method of puncture surgery robot based on optical navigation was proposed in this paper. Puncture tool with infrared ball is used to determine the position of the needle tip by using the principle of spherical center fitting. Then the singular value decomposition method of the given same position is used to control the movement of the robot arm and establish the hand-eye calibration equation matrix. Finally, tool calibration transfer matrix is derived by combining the known equations, and experiments are designed to verify the system. Experiments show the positioning accuracy of the puncture surgery robot system is controlled within 1mm. And the calibration process only needs to control the robot arm to move three times and record coordinate information. Calibration time is controlled at about 1 minute.

5. References

[1] Kathryn G C, Tim F and Mehran A 2016 Medical Robotics & Computer Assisted Surgery **12** 3
[2] Zhou Z, Wu B, Duan J, Zhang X, Zhang N and Liang Z 2017 Journal of Biomedical Optics **22** 6
[3] Zhang Z, Zhang L and Yang G Z 2017 Computer Assisted Radiology and Surgery **12** 10
[4] Jiang G W, Luo M Z, Lu L J, Bai K Q, Abdelaziz O and Chen S 2018 Applied Optics **57** 8385
[5] Yang G Z, James C, Kevin C, Eric D, James D, Pierre E, Nobuhiko H 2017 Science Robotics **2** 4
[6] Zeng B W, Meng F L, Ding H, Liu W B, Wu D 2017 Journal of Biomedical Engineering **34** 2
[7] Liu G F, Yu X, Li C, Li G, Zhang X H and Li L Y 2016 Computer Assisted Surgery **21** 55
[8] Niu J D, Luan N and Gui H J 2019 Machine Design & Research **35** 3
[9] Arun K S 1987 IEEE transactions on pattern analysis and machine intelligence pami-9 5

Acknowledgements

Research supported by the Key Research and Development project of Shandong Province (2019GSF108032), Major Program of Shandong Province Natural Science Foundation (ZR2018ZC0437) and Key Research and Development Program of Shandong Province (2016ZDJS02A07).

Analytical Calculation of Large Deflection of Annular Thin Plate of Damping Valve

Yijie Chen[1],Wei Nie[1,2],Xiaodong Gao[1], Baoqiang Li[1],Yiqiang Wan[1]

[1]China North Vehicle Research Institute, Beijing
[2] School of Mechanical Engineering, Beijing Institute of Technology, Beijing

chenyijie1206@163.com

Abstract: Aiming at the phenomenon of bending deformation caused by the pressure difference of the throttle plate in the working process of the damping valve, the analytical calculation of the large deflection of the annular thin plate was carried out. On the basis of extracting the physical model of the thin plate, the boundary conditions and initial conditions of the annular thin plate were set by using the Qian's perturbation method, and the nonlinear large deflection deformation analytical formula was derived; Through the finite element modeling, the accurate numerical solution of thin plate deformation was obtained, and the key parameters of the analytical formula were corrected by the least squares method. Through verification, the modified analytical formula had the characteristics of easy calculation and high calculation accuracy, and the influence law of the large deflection of the throttle valve was studied, which provided the necessary theoretical support for the accurate design of the damping valve performance.

1. Introduction

In the vehicle suspension damping device, there are many ways to generate the damping force, and the use of the throttle plate deformation slit to throttle the oil is widely used in engineering practice. After the valve is opened, the throttle plate is mainly in a large deflection state. However, in the existing design method, the small deflection deformation based on the classical loading method is mostly solved, and such an approximation method causes a large error. It is difficult to make the actual damping force of the damper correspond to the theoretical value, which does not meet the design requirements well[1-2].

Due to the complicated boundary conditions of the large deflection of the annular thin plate, the research results are few, so there is no mature theory applicable to engineering practice. Based on this situation, this paper proposes a derivation and systematic study on the large deflection of the annular throttle valve using the curve fitting method combining the Qian's perturbation method and the finite element analysis.

2. Physical model of throttle plate

Figure 1 is a schematic view showing the assembly of the throttle plate of the vibration damping device, the throttle valve piece is installed in the end of the piston. During the reciprocating motion of the piston, due to the different pressures of the upper and lower chambers, the pressure difference generated causes the outer edge of the valve plate to deform, and the oil will flow through the deformed annular gap, thereby generating a damping force to attenuate the vibration of the vehicle body. The throttle plate is separately extracted for simplification of the model, as shown in figure 2, in which the inner ring of the annular valve is fully constrained and the outer ring is in a freely deformed state. In the figure, q is the uniform load, h is the thickness of the valve, r_a is the inner radius of the valve, and r_b is the outer radius of the valve.

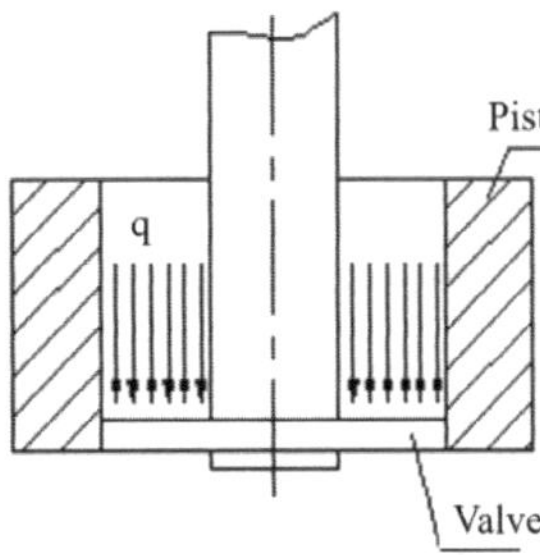

Figure 1.Shock absorber damping valve installation diagram

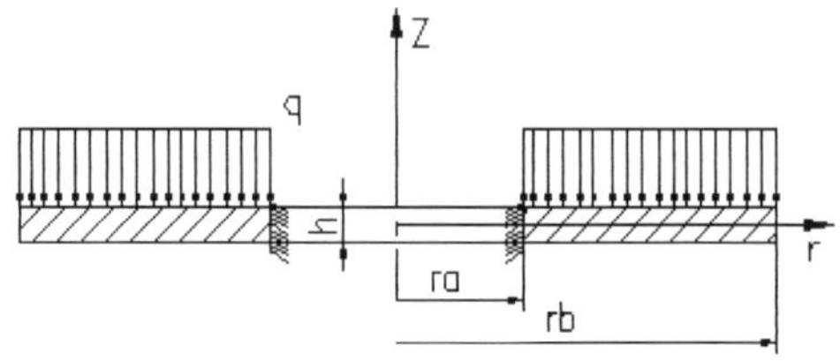

Figure 2. Throttle valve force model

3. Derivation of large deflection deformation of throttle plate

3.1 Derivation of deflection deformation based on perturbation method

The Von Kanman axisymmetric thin plate equation is transformed into the following form[3-4]:

$$\begin{cases} y^2 \dfrac{d^2\varphi(y)}{dy^2} = \varphi(y)S(y) + y^2\psi(y) \\ y^2 \dfrac{d^2S(y)}{dy^2} = -\dfrac{1}{2}\varphi^2(y) \end{cases} \tag{1}$$

Introduce the following dimensionless parameters:

$$\begin{cases} y = \dfrac{r^2}{r_b^2};\ W = \sqrt{3(1-u^2)}\dfrac{w}{h};\ \varphi(y) = y\dfrac{dW}{dy}; \alpha = \left(\dfrac{r_a}{r_b}\right)^2 \\ S(y) = 3(1-u^2)\dfrac{r_b^2 N_r y}{Eh^3};\ \psi(y) = \dfrac{3r_b^2(1-u^2)\sqrt{3(1-u^2)}}{4\pi Eh^4 y}R; \end{cases} \tag{2}$$

Where w is the deformation of the outer edge of the annular valve piece, u is the Poisson's ratio, E is the elastic modulus, N_r is the radial film tension, and R is the combined external load. In the case of uniform load, there is:

$$R = \int_{r_a}^{r} q(r)2\pi r dr = r_b^2\pi\int_{\alpha}^{y} q dy \tag{3}$$

Bring equation (3) into (2) and organize it:

$$\psi(y) = \frac{3r_b^4(1-u^2)\sqrt{3(1-u^2)}}{4Eh^4}q\left(1-\frac{\alpha}{y}\right) = Q\left(1-\frac{\alpha}{y}\right) \tag{4}$$

Give the boundary conditions of the physical model of the valve plate in Figure 2:

IOP Conf. Series: Materials Science and Engineering **717** (2020) 012009 doi:10.1088/1757-899X/717/1/012009

$$y=\alpha:\begin{cases}\varphi(y)=0\\ S(y)=\dfrac{2\alpha}{1+u}\dfrac{dS(y)}{dy}\end{cases}\qquad y=1:\begin{cases}\varphi(y)=\dfrac{2}{1-u}\dfrac{d\varphi(y)}{dy}\\ S(y)=0\end{cases}\tag{5}$$

Bring into (1) and organize it:

$$\begin{cases}y^2\dfrac{d^2\varphi(y)}{dy^2}=\varphi(y)S(y)+y^2Q\left(1-\dfrac{\alpha}{y}\right)\\ y^2\dfrac{d^2S(y)}{dy^2}=-\dfrac{1}{2}\varphi^2(y)\end{cases}\qquad \alpha<y<1\tag{6}$$

Let the dimensionless deflection function $W(y)$ is[5]:

$$W(y)=\int_{\alpha}^{y}\frac{1}{\xi}\varphi(\xi)d\xi\tag{7}$$

Then take the dimensionless deflection function $W_m(y)=\int_{\alpha}^{1}\frac{1}{\xi}\varphi(\xi)d\xi$ at the outer edge $(y=1)$ of the annular valve plate as the perturbation parameter.

Expand all unknown functions into power series of $W_m(y)$:

$$W(y)=\sum_{i=1}^{\infty}W_i(y)W_m^{2i-1}\qquad \varphi(y)=\sum_{i=1}^{\infty}\varphi_i(y)W_m^{2i-1}\tag{8}$$

$$S(y)=\sum_{i=1}^{\infty}S_i(y)W_m^{2i}\qquad Q=\sum_{i=1}^{\infty}\alpha_iW_m^{2i-1}\tag{9}$$

Where Q is the dimensionless parameter of the load strength, $W_i(y),\varphi_i(y),S_i(y)$ and α_i are the undetermined coefficients, and i is the order of the solution.

Under normal circumstances, when i=2, the accuracy requirement can be satisfied, and equations (8) and (9) are respectively brought into equations (5) and (6) and are arranged:

$$\begin{cases}\varphi_1(\alpha)W_m+\varphi_2(\alpha)W_m^{\ 3}=0\\ S_1(\alpha)W_m^{\ 2}+S_2(\alpha)W_m^{\ 4}=\dfrac{2\alpha}{1+u}\left(S_1'(\alpha)W_m^{\ 2}+S_2'(\alpha)W_m^{\ 4}\right)\end{cases}\tag{10}$$

$$\begin{cases}\varphi_1(1)W_m+\varphi_2(1)W_m^{\ 3}=\dfrac{2}{1-u}\left(\varphi_1'(1)W_m+\varphi_2'(1)W_m^{\ 3}\right)\\ S_1(1)W_m^{\ 2}+S_2(1)W_m^{\ 4}=0\end{cases}\tag{11}$$

$$\begin{cases} y^2\varphi_1''(y)W_m + y^2\varphi_2''(y)W_m^{\ 3} = \varphi_1(y)S_1(y)W_m^{\ 3} + \varphi_1(y)S_2(y)W_m^{\ 5} + \varphi_2(y)S_1(y)W_m^{\ 5} \\ +\varphi_2(y)S_2(y)W_m^{\ 7} + y^2\left(\alpha_1 W_m + \alpha_2 W_m^{\ 3}\right)\left(1-\dfrac{\alpha}{y}\right) \\ y^2 S_1''(y)W_m^{\ 2} + y^2 S_2''(y)W_m^{\ 4} = -\dfrac{1}{2}\left(\varphi_1^2(y)W_m^{\ 2} + \varphi_2^2(y)W_m^{\ 6} + 2\varphi_1(y)\varphi_2(y)W_m^{\ 4}\right) \end{cases} \tag{12}$$

Compare the coefficients of the similar terms at both ends of the equations (10) to (12) and sort them out:

$$W_m:\begin{cases} \varphi_1'' = \alpha_1\left(1-\dfrac{\alpha}{y}\right) \\ \varphi_1(\alpha)=0, \varphi_1(1)=\dfrac{2}{1-u}\varphi_1'(1) \end{cases} \qquad W_m^{\ 2}:\begin{cases} y^2 S_1'' = -\dfrac{1}{2}\varphi_1^{\ 2} \\ S_1(\alpha)=\dfrac{2\alpha}{1+u}S_1'(\alpha), S_1(1)=0 \end{cases} \tag{13}$$

$$W_m^{\ 3}:\begin{cases} y^2\varphi_2'' = \varphi_1 S_1 + y^2\alpha_2\left(1-\dfrac{\alpha}{y}\right) \\ \varphi_2(\alpha)=0, \varphi_2(1)=\dfrac{2}{1-u}\varphi_2'(1) \end{cases} \qquad W_m^{\ 4}:\begin{cases} y^2 S_2'' = -\varphi_1\varphi_2 \\ S_2(\alpha)=\dfrac{2\alpha}{1+u}S_2'(\alpha), S_2(1)=0 \end{cases} \tag{14}$$

From equations (7) and (8) there are:

$$\int_\alpha^1 \frac{1}{\xi}\varphi(\xi)d\xi = \int_\alpha^1 \frac{1}{\xi}\varphi_1(\xi)d\xi * \int_\alpha^1 \frac{1}{\xi}\varphi(\xi)d\xi + \int_\alpha^1 \frac{1}{\xi}\varphi_2(\xi)d\xi * \left(\int_\alpha^1 \frac{1}{\xi}\varphi(\xi)d\xi\right)^3 \tag{15}$$

Compare the similar items at both ends:

$$\int_\alpha^1 \frac{1}{\xi}\varphi_1(\xi)d\xi = 1 \qquad \int_\alpha^1 \frac{1}{\xi}\varphi_2(\xi)d\xi = 0 \tag{16}$$

The simultaneous equations (13) to (16) can be used to find the expressions of α_1、α_2、φ_1、φ_2 and S_1 respectively, and bring them into the following equation[6]:

$$Q = \alpha_1 W_m + \alpha_2 W_m^{\ 3} \tag{17}$$

And finishing:

$$r_b^4 q = 4Ehw\left[\frac{\alpha_1 h^2}{3\left(1-u^2\right)} + \alpha_2 w^2\right] \tag{18}$$

The above formula is a large deflection function expression of the second-order annular thin plate under uniform load obtained by the Qian's perturbation method.

3.2 curve fitting analytical formula

Formulate the linear function expression of equation (18):

$$w = \frac{r_b^4 q - 4E\alpha_2 h w^3}{16\alpha_1 D} \tag{19}$$

Where D is the bending stiffness of the valve plate.
Write the residual error squared sum expression of the flex function:

$$Q=\sum_{i=1}^{n}(w-w_i)^2=\sum_{i=1}^{n}\left(\frac{r_b^4 q_i-4E\alpha_2 h_i w_i^3}{16\alpha_1 D}-w_i\right)^2 \tag{20}$$

Where $i(1\cdots n)$ is the number of data sets, h_i is the thickness of the valve, q_i is the load value in the different data sets, and w_i is the deflection value obtained from the finite element.
Solve the partial derivatives of the above equations α_1 and α_2 respectively:

$$\frac{\partial Q}{\partial \alpha_1}=-2\sum_{i=1}^{n}\left[\frac{r_b^4 q_i-4E\alpha_2 h_i w_i^3}{16D\alpha_1^2}\left(\frac{r_b^4 q_i-4E\alpha_2 h_i w_i^3}{16\alpha_1 D}-w_i\right)\right]=0 \tag{21}$$

$$\frac{\partial Q}{\partial \alpha_2}=-2\sum_{i=1}^{n}\left[\frac{Eh_i w_i^3}{4\alpha_1 D}\left(\frac{r_b^4 q_i-4E\alpha_2 h_i w_i^3}{16\alpha_1 D}-w_i\right)\right]=0 \tag{22}$$

For the simultaneous (22) and (23), $\alpha_1=48.28, \alpha_2=1.03$ can be obtained. The expression of the large deflection function of the annular valve based on curve fitting is:

$$r_b^4 q=Ehw\left[64.38\frac{h^2}{(1-u^2)}+4.14w^2\right] \tag{23}$$

4 Analytical verification and analysis

Using curve fitting analytical equation (23) and finite element to calculate the large deflection of the outer edge of the throttle plate under different loads, the comparison results are shown in figure 3- 4 by Matlab.

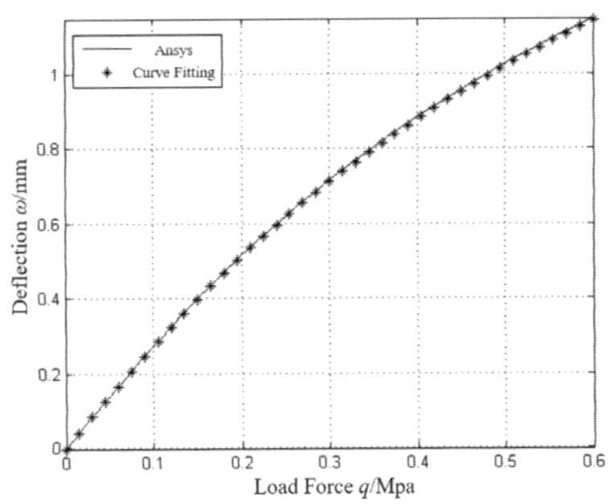

Figuge 3. The contrastive results when slice thickness is *0.4mm*

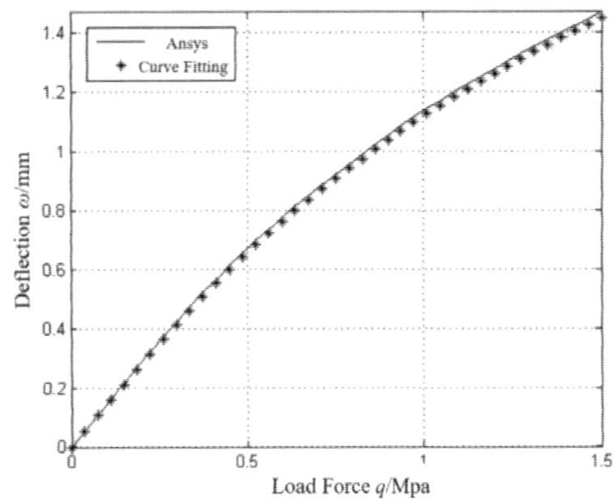

Figure 4. The contrastive results when slice thickness is *0.5mm*

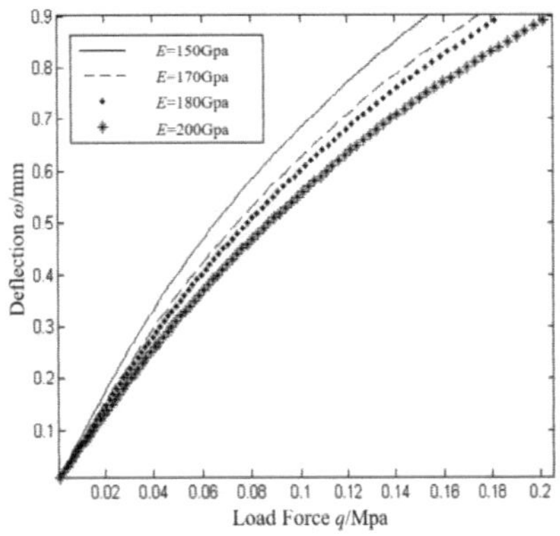

Figure 5. The contrastive result in different slice E

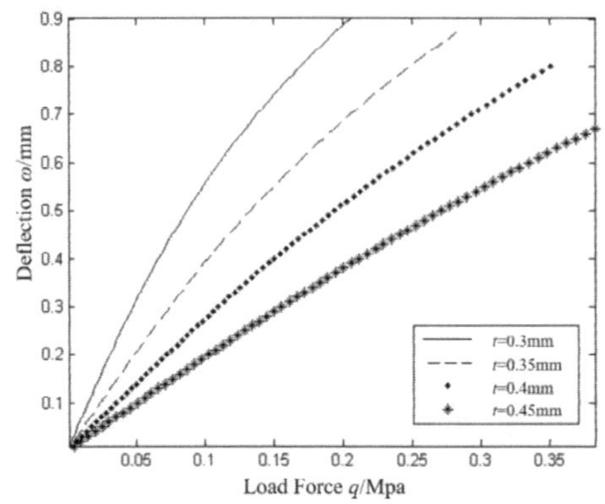

Figure 6. The contrastive result in different slice thickness

It can be seen from figure 5-6 that with the increase of the elastic modulus, the large deflection of the outer edge of the annular valve plate under the same load is gradually reduced; as the thickness increases, the outer edge of the annular valve plate is greatly deflected, the amount of deformation is reduced.

5. Conclusions

According to the least squares principle fitting, the relevant parameters in the analytical formula are determined. The correctness and accuracy of the fitting analytical formula are verified. The influence of the structural parameters and material properties of the valve on the large deflection is analyzed.

The analytic formula and simulation conclusion of the large deflection of the throttle plate obtained by curve fitting provide the possibility of accurate design of the damping valve, and provide an effective calculation method for the independent research and development of the products.

6. References

[1] Nieto A J, Morales A L, Trapero J R, et al. 2011 An adaptive pneumatic suspension based on the estimation of the excitation frequency. *Journal of Sound and Vibration*, **330(9)** pp 1891-1903.

[2] Morales A L, Nieto A J, Chicharro J M, et al. 2015 An adaptive pneumatic system for the attenuation of random vibrations. *Journal of Vibration and Control*, **21(5)** pp 907-918.

[3] Pu H Y, Luo X, Chen X D. 2011 Modeling and analysis of dual-chamber pneumatic spring with adjustable damping for precision vibration isolation. *Journal of Sound and Vibration*, **330(15)** pp 3578-3590.

[4] Li Z L, Li X Q, Chen X D. 2014 Generic vibration criteria-based dual-chamber pneumatic spring vibration isolation table design. *P I Mech Eng B-J Eng*, **228(12)** pp 1621-1629.

[5] Ryaboy V M. 2015 Static and dynamic stability of pneumatic vibration isolators and systems of isolators. *Journal of Sound and Vibration*, **333(1)** pp 31-51.

[6] Chai T, Dreyer J T, Singh R. 2015 Nonlinear dynamic properties of hydraulic suspension bushing with emphasis on the flow passage characteristics. *P I Mech Eng D-J Aut*, **229(10)** pp 1327-1344.

Effects of Manufacturing Tolerances on Dynamic Properties of Journal Bearings

Xun Ma[1,3], Wubin Xu[1,2]*, Sun Jin[1,3], Xueping Zhang[1,3] and Bing Li[2]

[1]Shanghai Key Laboratory of Digital Manufacture for Thin-walled Structures, Shanghai Jiao Tong University, China
[2]Faculty of Mechanical Engineering, Guangxi University of Science and Technology, China
3State Key Laboratory of Mechanical System and Vibration, Shanghai Jiao Tong University, China

Email: xuwub@163.com

Abstract. Dynamic properties of journal bearings were analysed when manufacturing tolerances existed. A methodology was proposed to predict journal bearing properties and to seek a relationship between the manufacturing tolerances and the dynamic coefficients. All the dynamic coefficients of journal bearings were investigated against Sommerfeld number and four types of standard tolerance grades in laminar flow regimes respectively. The results showed that manufacturing tolerances had a significant influence on the dynamic coefficients, especially the cross terms of both stiffness and damping coefficients. The more the magnitude of manufacturing tolerance grade is, the more variations of dynamic coefficients are. It is recommended to take the effect of tolerances on the dynamic performance of journal bearings on its design stage.

1. Introduction

In most of mechanical structures, manufacturing tolerances have nearly no or tiny influences on the static or dynamic properties of these mechanisms. However, it is not always the case for the rotor supported with hydrodynamic journal bearings. It was reported that manufacturing tolerances could have significant effects on the performance of journal bearings[1-3]. The performance of the rotor bearing system with the affected journal bearings could vary significantly compared to the non-affected rotor system, such as undergoing a process from a stable state to a sub-synchronous, synchronous or super-synchronous vibration. Therefore, it is necessary to investigate how the manufacturing tolerances affect the performance of journal bearings and the rotor which the journal bearings support.

There were limited works on the relationship between manufacturing tolerances and the dynamic properties of journal bearings. Pande and Somasundaram[4] studied the effect of manufacturing tolerances on the performance of aerostatic journal bearings using a numerical method. The manufacturing tolerances included ovality, lobbing, cylindricity and surface roughness. Hashimoto *et al.*[5, 6] investigated the steady and dynamic performance of a worn journal bearing. The results showed that the worn effect could have a significant difference on the dynamic properties of worn journal bearings. Although the worn effect was different from tolerances, it affected the bearing surface in the same way with manufacturing tolerances. Vijayaraghavan *et al.*[7] and Iwamoto and Tanaka[8] studied the influence of out of roundness on the bearing bush using a numerical method, but the research object was limited on the bearing bush and the tolerance investigated was just the out of roundness of the bearing bush. Fillon and Waldemar *et al.*[9, 10] investigated the effect of manufacturing tolerances on the static and dynamic performance of tilting pad journal bearings using a numerical method. Although the type of bearings investigated was tilting pad journal bearings, the research approach proposed was illuminating and worth to be learnt. Xu *et al.*[11, 12] presented a theoretical method to study the effect of out of roundness of the journal on the performance of rotor bearing systems. An experimental rig was built in order to verify the theoretical results and a good agreement was observed. All the former works admitted a similar methodology that using one or one set of regular curves to form the journal or the bearing surface, which lies in the range of manufacturing tolerances. However, in real physical systems, the surface of a journal or a bearing liner varies randomly in the tolerances limit, which makes a difference between results of theoretical prediction and real surface shape. For a journal bearing, the minimum film thickness may be close to the tolerance grade specified to the bearing, which could affect the performance of the bearing severely. Therefore, manufacturing tolerances should be considered more comprehensively in order to foretell the bearing properties more accurately.

Based on the former analysis, it is necessary to consider the effect of manufacturing tolerances on the dynamic properties which directly affect the stability of the rotor bearing system supported by them. These coefficients could be evaluated by means of the infinitesimal approach proposed by Lund[13] or the finite perturbation method of Qiu and Tieu[14]. Many works studied the dynamic properties related to journal bearings with the help of the two approaches, and very good agreement is achieved either compared with each other or verified by the experimental results[2, 3, 9-12, 15-18]. In this work, the dynamic properties of journal bearings are analyzed when manufacturing tolerances exist. Firstly, the film pressure distribution of finite length bearing, influenced by manufacturing tolerances, is evaluated by a modified Reynolds equation using the Ng-Pan and Elrod model. The new derived Reynolds equation could be used to predict the static and dynamic performance of a journal bearing in ether laminar or turbulent flow regime. Secondly, the stiffness and damping coefficients are calculated using the infinitesimal method by means of representing the perturbation equation of Reynolds equation. A robust numerical algorithm is used to acquire the coefficients of journal bearings, and the coefficients are verified and compared with the results of former works.

2. Theory and Materials

2.1. *Tolerances and fundamental deviations*

In industry application, to achieve an exact dimension for a part is impossible due to the inherent inaccuracy of machine tools, various manufacturing processes, and other uncontrollable factors. The exact dimensions of a part are always determined by tolerances and fundamental deviations. Table 1 shows a few standard tolerance grades and fundamental deviation *g* for nominal dimensions of 25mm and 50mm according to ISO[19].

Table 1 Standard tolerance grades IT for nominal dimensions of 25mm and 50mm according to ISO[19]

Nominal	Tolerance Grade, (μm)				Fundamental Deviation, *g*
(mm)	IT5	IT6	IT7	IT8	(μm)
25	9	13	21	33	-7
50	11	16	25	39	-9

2.2. *Journal Bearing Model*

In this work, a numerical model of a journal bearing with manufacturing tolerances is established based on the parameters provided in Table 2. Because that the tolerances about the bearing liner has been studied thoroughly and completely by former works[9, 10, 20], and that the part surface are controlled by a few types of tolerances together, the following assumptions are made. Firstly, the bearing model focuses only on the tolerances of the shaft diameter. Secondly, the bearing liner is well manufactured without any geometrical errors. Thirdly, for simplicity and concentrating on the influence of tolerances, the thermal effect is not considered in the bearing model. Fourthly, the journal bearing studied is assembled by the hole-basis system of fits. Finally, tolerances about the bearing length are omitted because the bearing length has little influence on the dynamic properties of the journal bearing in the long bearings, according to Waldemar and Michel's work[9, 10], or to the parameters in Sommerfeld number.

Table 2 Parameters Used in Models of Journal Bearings

Parameters	Value
Nominal Journal Diameter D, mm	25
Bearing Diameter D_b, mm	25.04
Bearing Length L, mm	25
Nominal Clearance c, μm	20
Oil Viscosity μ, Pa·s	7.5×10^{-3}
Rotational Speed ω, rad/s	314.15
Shaft Diameter Tolerances	IT5, IT6, IT7, IT8
Shaft Diameter Deviation	g

Figure 1 shows a journal bearing with the tolerance and fundamental deviation specified to the shaft. In order to investigate the influence from the tolerance zone, three typical dimensions are selected to succeed the study, including nominal dimension, maximum dimension and minimum dimension. When the tolerance of the shaft diameter is selected, three typical dimensions are nominal shaft diameter, maximum shaft diameter and minimum diameter. These three rotor diameters are selected because of the fact that all the tolerances in journal bearings, including roundness, cylindricity, straightness, verticality etc, must obey the envelope requirement, which requires that the envelope of perfect form for the feature at maximum material size shall not be violated.

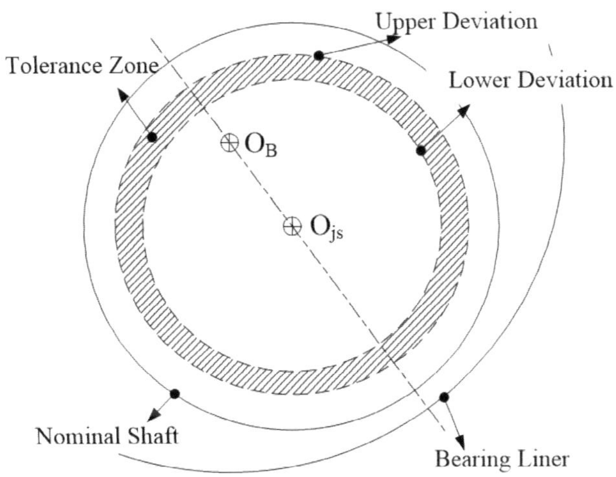

Figure 1 Schematic of a journal bearing with tolerance and deviation on the shaft.

2.3. Theoretical Model for Film Lubrication

The Reynolds equation for laminar and turbulent fluid is summarized and cited by a lot of former works[16, 21-23], and could be conveyed as

$$\frac{\partial}{\partial x}\left(\frac{h^3}{k_x\mu}\frac{\partial p}{\partial x}\right)+\frac{\partial}{\partial z}\left(\frac{h^3}{k_z\mu}\frac{\partial p}{\partial z}\right)=\frac{\omega R}{2}\frac{\partial h}{\partial x}+\frac{\partial h}{\partial t} \tag{1}$$

where R is shaft radius, μ is the lubricant oil viscosity, h is the film thickness, x is the circumferential position, z is the axial or longitudinal position, ω is rotational velocity of shaft, t is the operating time, k_x and k_z are the turbulence coefficients along circumferential and longitudinal directions, respectively.

Particularly, when the coefficients k_x and k_z are all set to 12, the equation (1) becomes a laminar regime equation.

In order to acquire a more robust and efficient numerical calculation program, the following parameters are defined to nondimensionalize the equation (1),

$$\bar{x}=\frac{x}{2\pi R},\ \bar{z}=\frac{z}{L},\ \bar{h}=\frac{h}{c},\ \bar{t}=\omega t,\ \bar{p}=\frac{p}{\mu\omega}\left(\frac{c}{R}\right)^2 \tag{2}$$

where L is the length of journal bearing, c is the clearance of journal bearing. After substitute the equation (2) into the Reynolds equation (1), the dimensionless Reynolds equation could be represented in a conservative form

$$\frac{\partial\bar{h}}{\partial\bar{t}}+\frac{1}{4\pi}\frac{\partial\bar{h}}{\partial\bar{x}}-\frac{\partial}{\partial\bar{x}}\left(\frac{1}{4\pi^2k_{\bar{x}}}\bar{h}^3\frac{\partial\bar{p}}{\partial\bar{x}}\right)-\frac{\partial}{\partial\bar{z}}\left(\frac{1}{4\left(L/D\right)^2k_{\bar{z}}}\bar{h}^3\frac{\partial\bar{p}}{\partial\bar{z}}\right)=0 \tag{3}$$

where D is the diameter of shaft.

Equation (3) is solved by the finite volume method, and in order to make the computer program be robust and efficient, the Approximate Factorization (AF) algorithm is applied[15, 24]. The Swift-Stieber boundary condition is employed and a finer mesh grid 100×50 is used. And the convergence criterion for the pressure is 1×10^{-7}.

2.4. Dynamic Coefficients and Perturbation equation

There are many papers and books dealing with the dynamic coefficients of journal bearings[14, 15, 25-27]. An efficient infinitesimal perturbation method is adopted to solve the dynamic coefficients for journal bearings in this research. Based on the equation (3), an infinitesimal perturbation method is presented.

Suppose that the equilibrium point of the bearing is $O_{js}(\xi_o, \eta_o)$, and at some instant t, the shaft center locates on a distance (Δx, Δz) away from O_{js} along x and y direction respectively. Using the Taylor expansion, the film thickness $\bar{h}$ and the pressure $\bar{p}$ could be expressed by

$$\bar{h}=\bar{h}_o+\Delta\bar{x}\cos\theta+\Delta\bar{z}\sin\theta+H.O.T\left(\Delta\bar{x},\Delta\bar{z}\right) \tag{4}$$

$$\bar{p}=\bar{p}_o+\frac{\partial\bar{p}_o}{\partial\bar{x}_j}\Delta x_j+H.O.T\left(\Delta\bar{x}_j\right) \tag{5}$$

where $\bar{x}_j$ (j = 1, 2, 3 and 4) represent for variables ($\bar{x}$, $\dot{\bar{x}}$, $\bar{z}$ and $\dot{\bar{z}}$). $\bar{h}_o$ and $\bar{p}_o$ are the film thickness and the pressure at the equilibrium state.

Substituting the equation (4) and the equation (5) into equation (3), neglecting all the second terms and higher, omitting the time terms and sorting out the equation with each term $\Delta\bar{x}_j$, the following equations are obtained

$$\begin{aligned}
&\Re\left(\bar{p}_o\right)=\frac{1}{4\pi}\frac{\partial\bar{h}_o}{\partial\bar{x}}\\
&\Re\left(\bar{p}_x\right)=\frac{1}{2}sin\theta-\frac{\partial}{\partial\bar{x}}\left(\frac{3\bar{h}_o^2cos\theta}{4\pi^2k_x}\frac{\partial\bar{p}_o}{\partial\bar{x}}\right)-\frac{\partial}{\partial\bar{z}}\left(\frac{3\bar{h}_o^2cos\theta}{4(L/D)^2k_z}\frac{\partial\bar{p}_o}{\partial\bar{z}}\right)\\
&\Re\left(\bar{p}_z\right)=-\frac{1}{2}cos\theta-\frac{\partial}{\partial\bar{x}}\left(\frac{3\bar{h}_o^2sin\theta}{4\pi^2k_x}\frac{\partial\bar{p}_o}{\partial\bar{x}}\right)-\frac{\partial}{\partial\bar{z}}\left(\frac{3\bar{h}_o^2sin\theta}{4(L/D)^2k_z}\frac{\partial\bar{p}_o}{\partial\bar{z}}\right)\\
&\Re\left(\bar{p}_{\dot{\bar{x}}}\right)=cos\theta\\
&\Re\left(\bar{p}_{\dot{\bar{z}}}\right)=sin\theta
\end{aligned} \tag{6}$$

where a differential operator is defined as

$$\Re(\ )=\frac{\partial}{\partial\bar{x}}\left(\frac{h_o^3}{4\pi^2 k_x}\frac{\partial(\ )}{\partial\bar{x}}\right)+\frac{\partial}{\partial\bar{z}}\left(\frac{\bar{h}_o^3}{4(L/D)^2 k_z}\frac{\partial(\ )}{\partial\bar{z}}\right) \tag{7}$$

After solving the pressure and their derivations by equation (6), all the coefficients could be foretold by

$$\begin{aligned}
\bar{K}_{xx}&=-\int_0^{\bar{x}_{cav}}\int_{-1/2}^{1/2}\bar{p}_{\bar{x}}\cos\theta d\bar{x}d\bar{z},\ \bar{K}_{yx}=-\int_0^{\bar{x}_{cav}}\int_{-1/2}^{1/2}\bar{p}_{\bar{x}}\sin\theta d\bar{x}d\bar{z}\\
\bar{K}_{xy}&=-\int_0^{\bar{x}_{cav}}\int_{-1/2}^{1/2}\bar{p}_{\bar{z}}\cos\theta d\bar{x}d\bar{z},\ \bar{K}_{yy}=-\int_0^{\bar{x}_{cav}}\int_{-1/2}^{1/2}\bar{p}_{\bar{z}}\sin\theta d\bar{x}d\bar{z}\\
\bar{C}_{xx}&=-\int_0^{\bar{x}_{cav}}\int_{-1/2}^{1/2}\bar{p}_{\dot{\bar{x}}}\cos\theta d\bar{x}d\bar{z},\ \bar{C}_{yx}=-\int_0^{\bar{x}_{cav}}\int_{-1/2}^{1/2}\bar{p}_{\dot{\bar{x}}}\sin\theta d\bar{x}d\bar{z}\\
\bar{C}_{xy}&=-\int_0^{\bar{x}_{cav}}\int_{-1/2}^{1/2}\bar{p}_{\dot{\bar{z}}}\cos\theta d\bar{x}d\bar{z},\ \bar{C}_{yy}=-\int_0^{\bar{x}_{cav}}\int_{-1/2}^{1/2}\bar{p}_{\dot{\bar{z}}}\sin\theta d\bar{x}d\bar{z}
\end{aligned} \tag{8}$$

To validate the numerical algorithm, the bearing model proposed by Mohammad was analyzed, and all the dynamic coefficients predicted using the current bearing model are compared with the Mohammad's work in Figure 2[16]. It is found that all the stiffness coefficients fit for the Mohammad's results very well, and that all the damping coefficients agree with them fairly well. It is noted that all the dimensionless coefficients in Figure 2 are calculated using the methodology proposed by Mohammad.

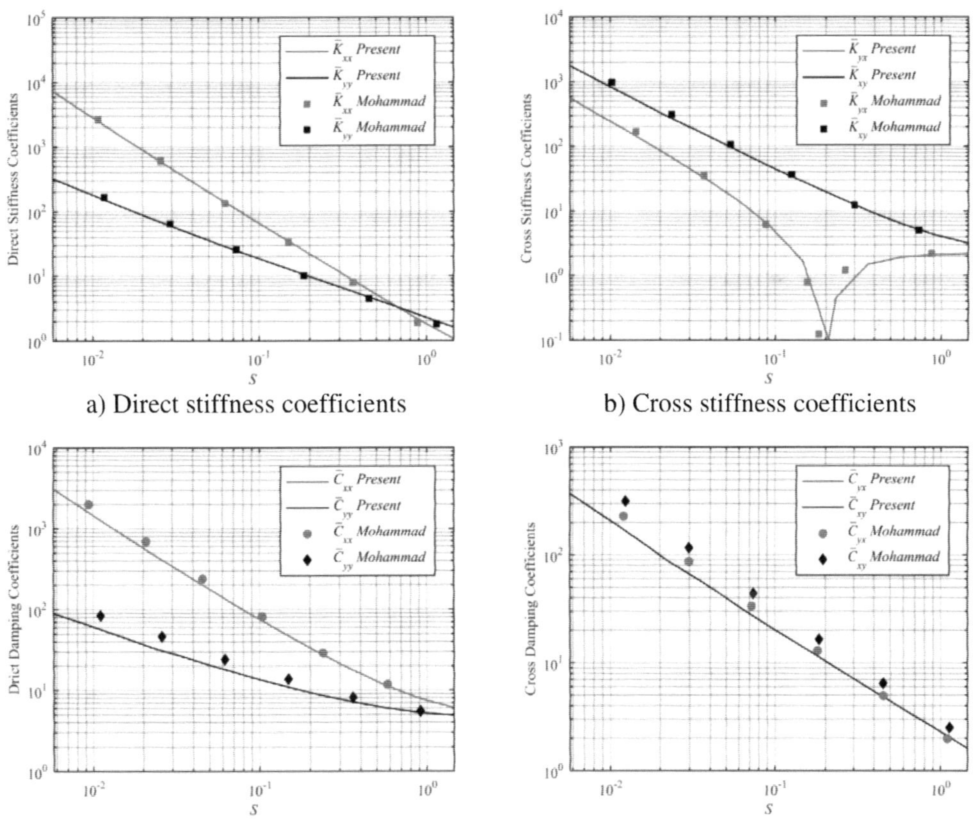

a) Direct stiffness coefficients

b) Cross stiffness coefficients

c) Direct damping coefficients

d) Cross stiffness coefficients

Figure 2 Comparison of dimensionless stiffness and damping coefficients with the present and Mohammad's results (L/D = 0.5).

3. Results and Discussion

The dynamic coefficients of journal bearings with standard tolerance grades from IT5 to IT8 are calculated with the help of Reynolds equation (3) and perturbation equation (6) and the parameters in

IOP Conf. Series: Materials Science and Engineering **717** (2020) 012010 doi:10.1088/1757-899X/717/1/012010

Table 2. The selected Sommerfeld number S used in the numerical model equals to 2.895, 1.628, 0.916, 0.515, 0.29, 0.163, 0.092, 0.051, 0.029, 0.016 and 0.009, respectively.

The variations of dimensionless dynamic coefficients of journal bearings are illustrated against selected standard tolerance grades in Figure 3 when Sommerfeld number S equals to 1.63 and fundamental deviation g is specified. It is found that all the dynamic coefficients vary as the standard tolerance grade increases from IT5 to IT8, the more the grade is, the more the variations of dynamic coefficients are. The coefficient least affected is the direct stiffness coefficient $\bar{K}_{xx}$ with tolerance grade IT5, fluctuating between 0.026 and 0.03 (corresponding to 13.9% about its nominal value 0.035). Therefore, it is benefit to limit the manufacturing tolerances to small tolerance grade for the purpose of stable operating.

Although Figure 3 shows information about the variations of dynamic coefficients as a few tolerance grades, a detailed relationship between the coefficients and Sommerfeld number is not illustrated clearly. For this purpose, the variations of the coefficients with the same tolerance grade (IT6) as Sommerfeld number S are represented in Figure 4 when fundamental deviation g is specified. As is shown in Figure 4, the curves of coefficients for the shaft with tolerance vary in an area confined by curves of lower limit and upper limit. It is found that the area surrounded by curves of lower limit and upper limit deviates from the nominal curve by a distance, the magnitude of this distance is determined by the deviation specified to the shaft.

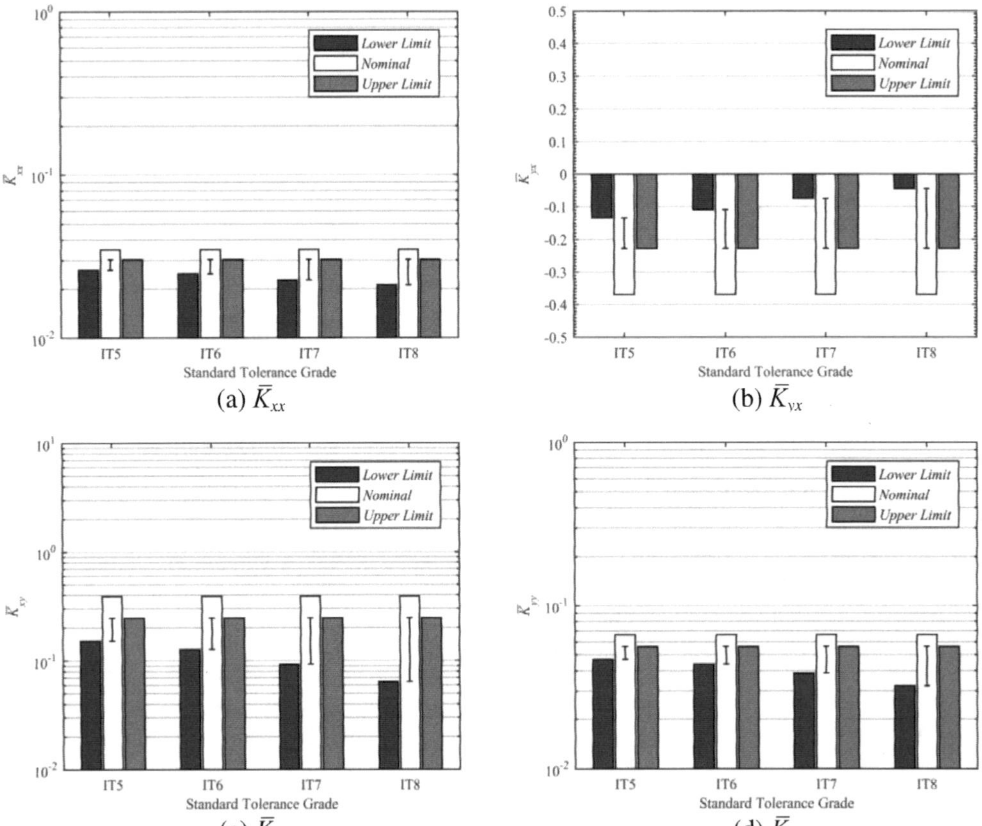

(a) $\bar{K}_{xx}$ (b) $\bar{K}_{yx}$

(c) $\bar{K}_{xy}$ (d) $\bar{K}_{yy}$

IOP Conf. Series: Materials Science and Engineering **717** (2020) 012010 doi:10.1088/1757-899X/717/1/012010

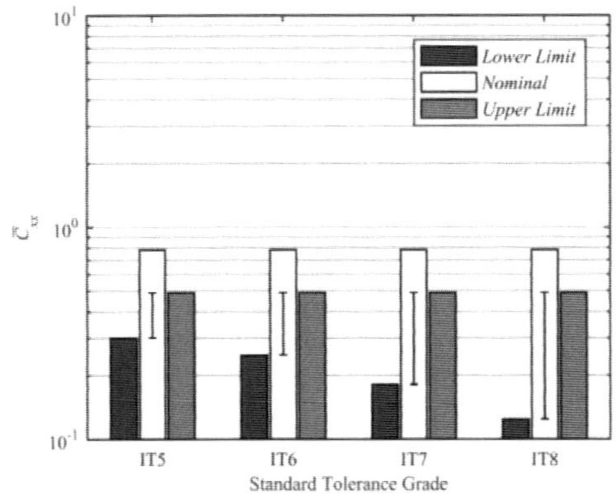

(e) $\overline{C}_{xx}$

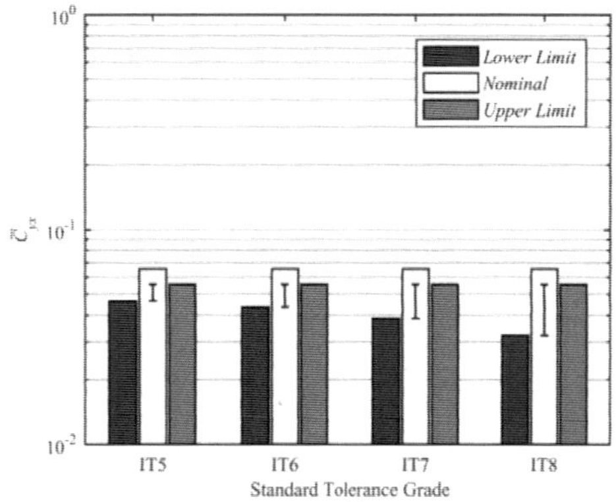

(f) $\overline{C}_{yx}$

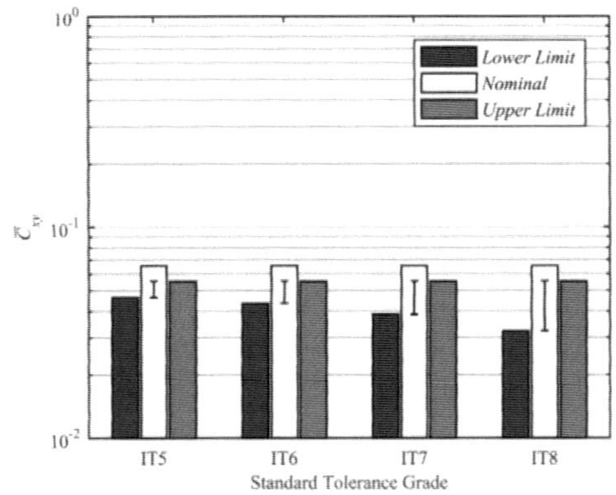

(g) $\overline{C}_{xy}$

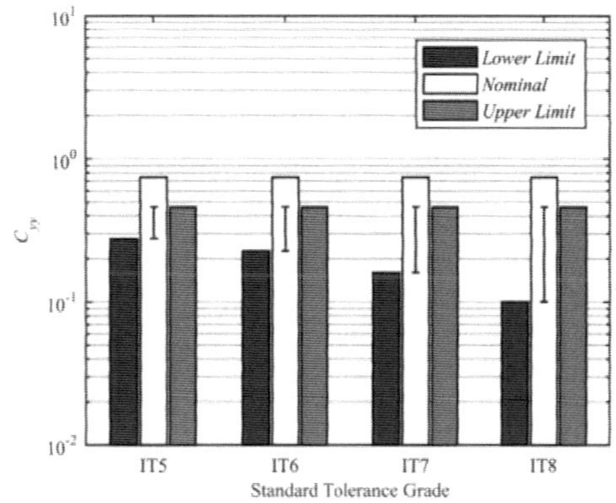

(h) $\overline{C}_{yy}$

Figure 3 Variations of the eight dynamic coefficients of journal bearings as the standard tolerance grades (S = 1.63, fundamental deviation: g).

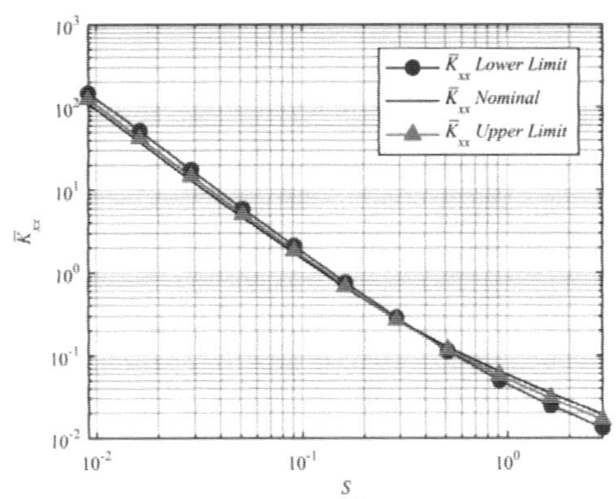

(a) $\overline{K}_{xx}$

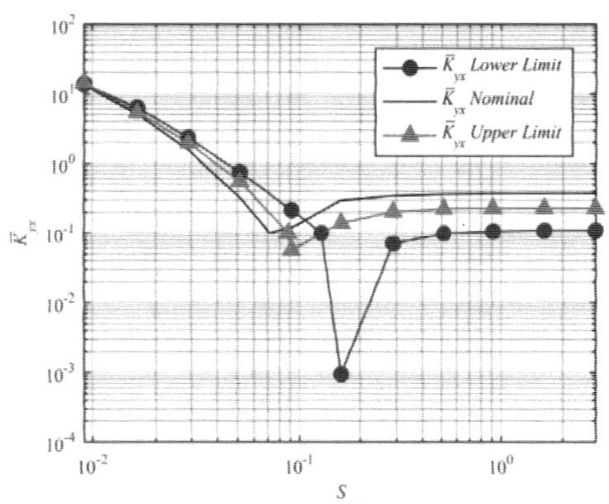

(b) $\overline{K}_{yx}$

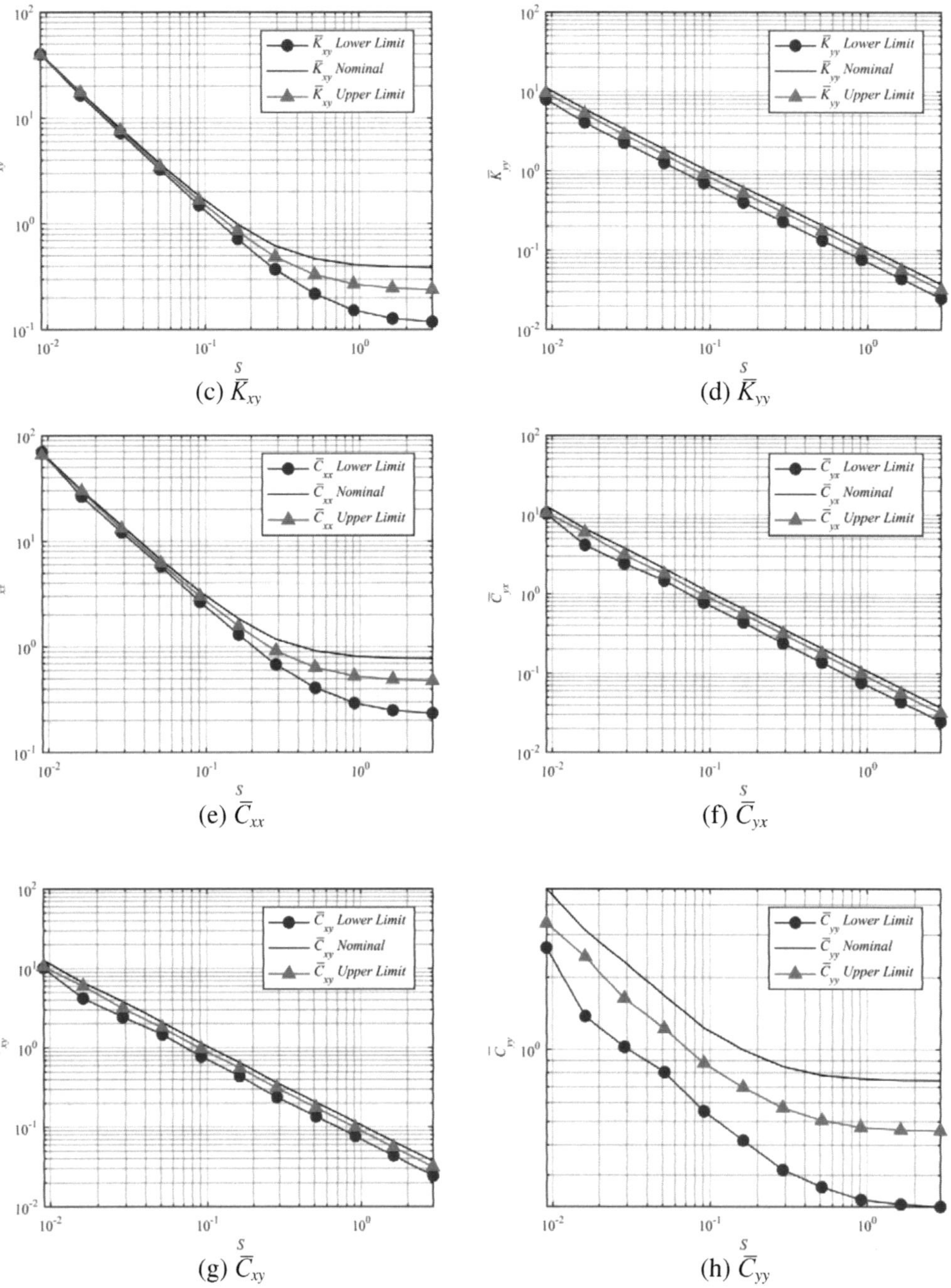

(c) $\overline{K}_{xy}$ (d) $\overline{K}_{yy}$

(e) $\overline{C}_{xx}$ (f) $\overline{C}_{yx}$

(g) $\overline{C}_{xy}$ (h) $\overline{C}_{yy}$

Figure 4 Variations of dynamic coefficients caused by manufacturing tolerances of diameter versus Sommerfeld number S (Standard tolerance grade: IT6, fundamental deviation: g).

4. Conclusion

In this work, the dynamic properties of journal bearings affected by manufacturing tolerances are analysed in laminar flow regime. For the efficiency and robustness of the numerical algorithm, the Reynolds equation and the corresponding disturbed equations for dynamic coefficients are nondimensionalized and represented using defined parameters. Four types of standard tolerance grades are introduced in the numerical model respectively, when fundamental deviation g is specified. It is found that manufacturing tolerances have a significant influence on the dynamic coefficients, especially the cross terms of both stiffness and damping coefficients. The more the magnitude of the tolerance grade is, the more variation of dynamic coefficients is. All the results could be used as a basis for foretelling the dynamic performance of journal bearings in the design stage of assigning manufacturing tolerances to journal bearings.

5. Reference

[1] Radford A and Fitzgeorge D 1977 *Wear.* **45** 311-22.

[2] Ogrodnik PJ, Xu W, Goodwin MJ and Bancroft GA 2011 *Proceedings of the Institution of Mechanical Engineers, Part J: Journal of Engineering Tribology.* **225** 1152-8.

[3] Ogrodnik PJ, Xu W, Goodwin MJ and Bancroft GA 2011 *Proceedings of the Institution of Mechanical Engineers, Part J: Journal of Engineering Tribology.* **225** 731-9.

[4] Pande S and Somasundaram S 1981 *Wear.* **66** 145–56.

[5] Hashimoto H, Wada S and Nojima K 1986 *ASLE Transactions.* **29** 565-71.

[6] Hashimoto H, Wada S and Nojima K 1986 *ASLE Transactions.* **29** 572-7.

[7] Vijayaraghavan D, Brewe DE and Keith JTG 1993 *Journal of Tribology.* **115** 538-43.

[8] Iwamoto K and Tanaka K 2005 *Tribology and Interface Engineering Series.* **Volume 48** 751-4.

[9] Fillon M, Dmochowski W and Dadouche A 2008 *Tribology Transactions.* **50** 387-400.

[10] Dmochowski W, Dadouche A and Fillon M 2008 *Tribology Transactions.* **51** 573-80.

[11] Xu W, Ogrodnik PJ, Li B, Zhang H and Gao S 2015 *Proceedings of the Institution of Mechanical Engineers, Part J: Journal of Engineering Tribology.* **229** 1347-58.

[12] Xu W, Ogrodnik PJ, Li B, Zhang H and Gao S 2015 *Proceedings of the Institution of Mechanical Engineers, Part J: Journal of Engineering Tribology.* **229** 1359-71.

[13] Lund JW and Thomsen KK 1978 *Topics in Fluid Film Bearing and Rotor Bearing System Design and Optimization, ASME.* 1-28.

[14] Qiu ZL and Tieu AK 1996 *Tribology Transactions.* **39** 469-75.

[15] Miraskari M, Hemmati F and Gadala MS 2018 *Journal of Tribology-Transactions of the Asme.* **140** 19.

[16] Miraskari M, Hemmati F, Alqaradawi MY and Gadala MS 2017 *Proceedings of the Institution of Mechanical Engineers, Part J: Journal of Engineering Tribology.* **231** 1254-67.

[17] Xu W, Ogrodnik PJ, Goodwin MJ and Bancroft GA 2009. 164-7.

[18] Ogrodnik PJ, Goodwin MJ, Bancroft GA and Xu W 2009. 168-71.

[19] ISO-286. ISO System of Limits and Fits—Part 1: Bases of Tolerances, Deviations and Fits. *International Standard.* 1988.

[20] Jones GJ and Martin FA 2008 *A S L E Transactions.* **22** 227-44.

[21] Taylor CM and Dowson D 1974 *Journal of Lubrication Technology.* **96** 36-46.

[22] Elrod JHG and Ng CW 1967 *Journal of Lubrication Technology.* **89** 346-62.

[23] Constantinescu VN 1959 *Proceedings of the Institution of Mechanical Engineers.* **173** 881-900.

[24] Vijayaraghavan D and Keith JTG 1990 *Journal of Tribology.* **112** 44-51.

[25] Lund JW and Saibel E 1967 *Journal of Engineering for Industry.* **89** 813-23.

[26] Lund JW 1964 *A S L E Transactions.* **7** 342-52.

[27] Lund JW 1974 *Journal of Engineering for Industry.* **96** 509-17.

Funding

We gratefully acknowledge the financial supports from the National Natural Science Foundation of China (NSFC) under the project No.51265004, No.50865001, No.51105084, and Guangxi Natural Science Foundation under the project No.2011GXNSFF018004.

IOP Conf. Series: Materials Science and Engineering **717** (2020) 012010 doi:10.1088/1757-899X/717/1/012010

Appendix

Notation

c	clearance of journal bearing (m)
C_{ij}	damping coefficients (N s/m)
$\bar{C}_{ij}$	dimensionless damping coefficients, $\bar{C}_{ij} = C_{ij}c/\pi\mu LD(R/c)^2$
e	eccentricity (m)
D	nominal diameter of journal (m)
D_b	bearing diameter (m)
h	film thickness (m)
$\bar{h}$	dimensionless film thickness, $\bar{h} = h/c$
K_{ij}	stiffness coefficients (N/m)
$\bar{K}_{ij}$	dimensionless stiffness coefficients, $\bar{K}_{ij} = K_{ij}c/\pi\mu\omega LD(R/c)^2$
k_x, k_z	turbulent coefficients along circumferential and longitude direction
L	nominal length of journal bearing (m)
O_{js}	center of journal shaft in static operations
p	film pressure (Pa)
$\bar{p}$	dimensionless film pressure, $\bar{p} = p/\mu\omega(R/c)^2$
$\bar{p}_o$	dimensionless film pressure at static state
R	shaft radius (m)
S	Summerfeld numbers, $S = LD\mu N(R/c)^2/W$
t	operating time (s)
$\bar{t}$	dimensionless operating time, $\bar{t} = \omega t$
x	circumferential position specified to a journal bearing
$\bar{x}$	dimensionless circumferential position, $\bar{x} = x/2\pi R$
z	longitude position specified to a journal bearing
$\bar{z}$	dimensionless longitude position, $\bar{z} = z/L$
ω	shaft rotational speed (rpm)
μ	oil viscosity (Pa s)
(ξ_o, η_o)	equilibrium point of journal bearings in xz coordinate system

IOP Conf. Series: Materials Science and Engineering **717** (2020) 012011 doi:10.1088/1757-899X/717/1/012011

Linear Bearing Fault Detection Using an Artificial Neural Network Based on a PI Servo System with the Observer for High-speed Automation Machine

T Wanglomklang[1], P Chommaungpuck[1, 2] and J Srisertpol1*

[1] School of Mechanical Engineering, Suranaree University of Technology, Nakhon Ratchasima Province, Thailand

[2] Western Digital (Thailand) Co.Ltd, Bang Pa-in Industrial Estate, Ayutthaya Province, Thailand

E-mail: jiraphon@sut.ac.th

Abstract. This research presents the novel approach for linear bearing fault detection by using Artificial Neural Network (ANN) based on observable information for high-speed automation machine. The dynamics modelling of feed drives and servo system design using pole placement technique were established to support the propose method. Three conditions of linear bearings which included healthy, 50 % of lubrication oil and starved lubrication were set up. Feature extraction of the data was analyzed by statistical approach. The results explains clearly that the control system design has a performance for tracking response and the ANN model can achieved 99.7 % accuracy by using the Levenberg Marquardt algorithm.

1. Introduction

Recently, Hard Disk Drive (HDD) has been the important data storage device. The Head Gimbal Assembly (HGA) is the major part in HDD. It's consist of two main components which are the slider and the suspension. The HGA uses for reading and writing the data. The process to produce HGA, the Auto Core Adhesion Mounting machine (ACAM) was used for dispend a glue and attaching the slider onto the suspension by utilizing the feed drive actuator for moving fixture clamp to the desired position. In order to facilitate the fluent motion and eliminate the effect of friction the linear bearing is an important supporting component in feed drive actuator, however, when the machine operating in continuous time and high speed condition the linear bearing easier to occur fault that affect to miss position of adhesive dispensing and slider attaching misalignment. In addition also decrease performance of the control system. The reviews on modern control design technique for feed drive actuator and linear bearing fault detection will be provided as follows. The comparison of PI-servo with Ogata method was studied by Saengsri. S. The experiments validate by robustness on external disturbance [1]. In addition, the electromagnetic levitation was applied the servo system using pole placement with state observer for controller design [2]. Xie D. Presented control system of feed drive in CNC machine by integrating the fuzzy logic with PID architecture for adjusting controller gain [3]. Friction model has extended to XY feed drive for improving the performance tracking with using the PID and state feedback control [4]. The ball bearing fault is considered as a critical element. Neural network classification using the vibration signal based on Hilbert footprint analysis was discussed. The results show propose method is given accuracy 87.3-100 % [5]. In addition, bearing fault detection employed the deep convolution neural network was presented by Zhao D. The method automatically recognizes in the time-frequency image and generate 98.3% accuracy [6]. In other hand, Chen Z. Studied on a rolling bearing fault diagnosis by providing deep neural network for classifying. Four feature applying into vibration signal in both time and frequency domain [7]. To study on fault detection and isolation of linear bearing in ACAM by using Artificial Neural Network (ANN). The data of motor current, FFT and crest factor was measured for the training data set. The experiments show that the model which training by use three parameters have a high accuracy at 93% [8]. Looking more closely though many fault detection techniques, most vibration signal was widely proposed, however, several machines are limitation area for installation and cost of sensor is one important to consider.

This paper presents an approach to modelling the dynamics of feed drives in high-speed automation machine such as ACAM and linear bearing fault detection based on observer information to replace the sensor measurement data. The mathematical model in the state space of servo motors with a lead screw is used to analyze in the modern control design technique. The design of PI servo system control using a pole placement method with state observer was established to support fault detection by using machine learning architecture. The experiment investigates the use of motor current, angular velocity and table velocity for feature extraction with statistical parameters to training ANN model. The proposed method has been tested with neural network toolbox in MATLAB software and the result was discussed with the highest accuracy of fault detection and diagnostics.

2. Dynamics modelling of feed drive with DC servo motor

The most common direct current (DC) servo motor is an actuator in feed drives because it can be used a wide range operating speed and driving with high torque for the machine tool. The feed drive system analyzed in this paper consist of a lead screw driven by a DC servo motor through the coupling. The clamping unit mounts on the worktable and supported by four linear bearing block and the linear guild to move axially. In ACAM, constrained of two XY stage and in each axis consist of the same component, therefor, design only one feed drive will be applied to others. The physical modelling of the single-axis is shown in Figure 1.

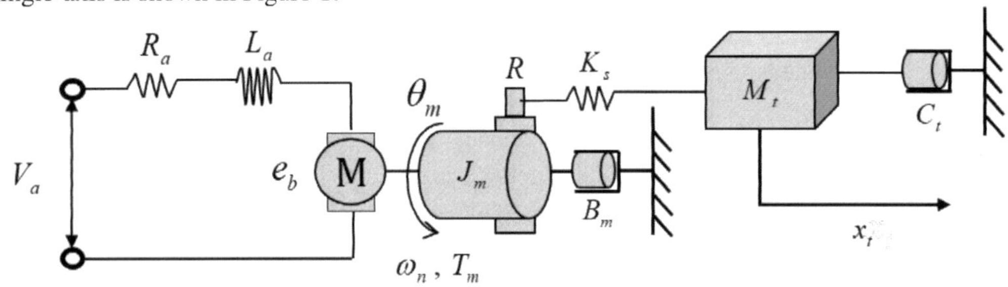

Figure 1. The physical modelling of feed drive

According to Figure 2 the state space model with state vector is arranged in matrix form show in equations (1) - (2).

$$\frac{d}{dt}\begin{bmatrix} i_a \\ \theta_m \\ \dot{\theta}_m \\ x_t \\ \dot{x}_t \end{bmatrix} = \begin{bmatrix} -\frac{R_a}{L_a} & 0 & -\frac{K_b}{L_a} & 0 & 0 \\ 0 & 0 & 1 & 0 & 0 \\ \frac{K_t}{J_m} & 0 & -\frac{B_m}{J_m} & 0 & 0 \\ 0 & 0 & 0 & 0 & 1 \\ 0 & \frac{RK_s}{M_t} & 0 & -\frac{K_s}{M_t} & -\frac{C_t}{M_t} \end{bmatrix} \begin{bmatrix} i_a \\ \theta_m \\ \dot{\theta}_m \\ x_t \\ \dot{x}_t \end{bmatrix} + \begin{bmatrix} \frac{1}{L_a} \\ 0 \\ 0 \\ 0 \\ 0 \end{bmatrix} u \tag{1}$$

$$y(t) = [0\ 0\ 0\ 1\ 0][i_a\ \theta_m\ \dot{\theta}_m\ x_t\ \dot{x}_t]^T \tag{2}$$

Where i_a is the motor armature current (A), θ_m is the motor rotation angle (rad) and x_t is the worktable position (m), J_m is the moment of inertia (kg·m2) and B_m is the coefficient of viscous friction (N·m·s/rad), R_a is the armature resistance (Ω), V_a is the armature voltage (V), L_a is armature inductance (H), K_t is the torque coefficient (N·m/A), K_b is the back electromotive force coefficient (V·s/rad), M_t regarded as total of worktable mass (kg) and C_t is the damping coefficient of the lead screw (N·s/m), *Ks* represents total equivalent stiffness coefficient of the lead screw (N/m) and *R* used as convert factor coefficient of motor rotation to lead screw position (mm/rad) its calculate by $l/2\pi$ which l is lead screw pith (mm). In this work the system identification approach has made practical use for estimating the system parameters of the feed drive system thought experimental data.

3. Fault detection architecture using artificial neural network with PI servo system

The design of the PI servo system is presented in this paper illustrated in Figure 2. In order to signify in transient and tracking response signal along with supporting the fault detection architecture.

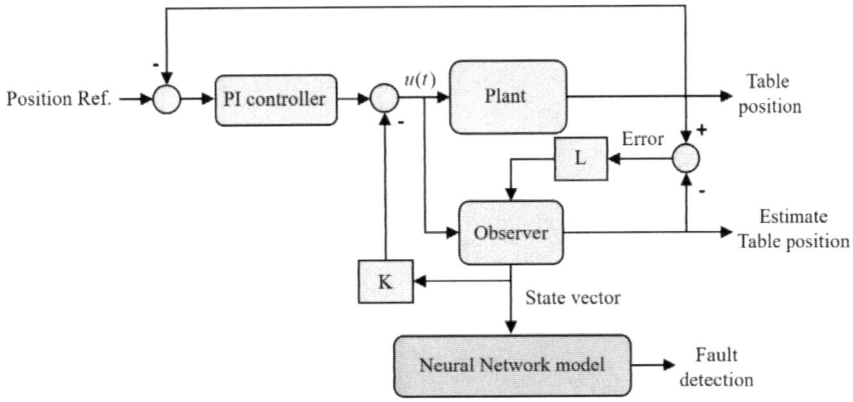

Figure 2. Fault detection scheme

3.1. PI servo system design with observer

The stability criteria are early precious to analyze, therefore, the pole placement method was applied to determine the state feedback gain and PI servo controller gain. The designing requires state variable feedback through gain *K*, in practices, the measuring of all state variables is difficult. The observer technique was used for estimating the state variable of the process under measurement output signal and measurable input. The first step of the control system design should be checking the controllability and the observability of the system. It was found that the matrix ranks equal to system order, which confirms the system is observable and controllable. The next step of the state feedback gain must be placing desire close loop pole because the location of the poles corresponds directly to the eigenvalues of the system, which control the characteristics of the response of the system. The design with position control of the worktable by providing over damp response. Based on state space equation as equation (3) all state vector was estimated by using observer that could be written as equation (4) - (5).

$$\begin{aligned} \dot{x} &= Ax + Bu \ , \quad x(t_0) = x_0 \\ y &= Cx \end{aligned} \tag{3}$$

$$\dot{\hat{x}} = A\hat{x} + Bu + L[y - C\hat{x}] \tag{4}$$

$$\dot{e} = [A - LC]e \tag{5}$$

From the equation (5) that is the dynamic behaviour of error between plant and observer. The matrices observer gain is determined from eigenvalue of matrix (A-LC) that meaning the observer pole which is should be faster than 10 times of close loop pole.

3.2. Data pre-processing and feature extraction

The process of data pre-processing is important for cleaning the measurement information before extraction the feature and model training. The state observer information includes current, table velocity and the estimation error was investigated on three bearing condition. The data collected to help of MATLAB/Simulink software in one operating speed and under control loop situation. Four statistical feature established on state vector data there are given as input to artificial neural network. To analyzed, difference number of hidden layer and three learning algorithm are used. The data separation is divided into training 70%, validation 15% and testing 15%. The discussion on feature extraction by implementing the four statistical parameters consist of standard deviation, kurtosis, crest factor and root mean square (RMS) are expressed below

$$\sigma = \sqrt{\frac{\sum_{i=1}^{n}(x_i - \bar{x})^2}{n-1}} \tag{6}$$

Where σ is the standard deviation, n is the number of samples, x_i is amplitude of individual samples and $\bar{x}$ is an average of samples.

$$Kurtosis = \frac{1}{n}\sum_{i=1}^{n}\left(\frac{x_i - \bar{x}}{\sigma}\right)^4 \tag{7}$$

The kurtosis is indicated the peak of wave form, basically, if the signal has a random with an impulsive the kurtosis coefficient is more than 3. The periodic waveform can be used crest factor to explain in equation (8).

$$Crest\ factor = \frac{peak\ value}{rms\ value} \tag{8}$$

Root Mean Square (RMS) is the power of the average value of the signal. As the amplitude increase the RMS value is mounted.

$$RMS = \sqrt{\frac{1}{n}\sum_{i=1}^{n} x_i^2} \tag{9}$$

3.3. Artificial Neural Network (ANN)

An artificial neural network arranged in a common tool of machine learning, which is spread in data science and engineering filed because now day computational calculation has become efficient for solving the numerical and mathematical problem. The ANN is a simulation architecture of the human biological nervous system as shown in Figure 3. The working process first accepts and weigh multiple inputs, then linear summation and sent to the mathematical activation function to make decisions.

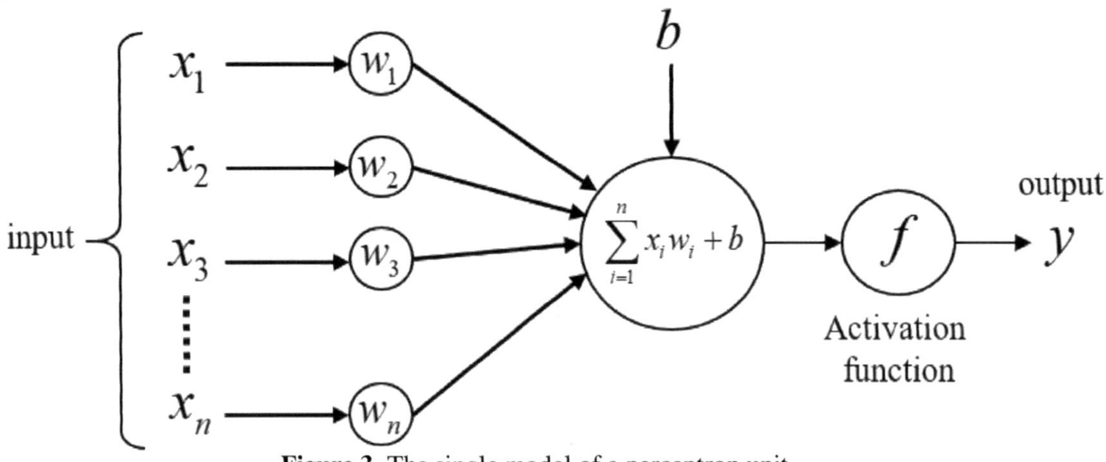

Figure 3. The single model of a perceptron unit

Within the model of ANN, amount of neural ordinarily impose into one or more layer perceptron. In definition the ANN consist of one input layer, one or multi of hidden layer and one output layer. The selecting of hidden layer number is huge impact to model performance, therefor in design phase many works should find the optimal number of hidden layers. The activation function usually implements relate in each type of problem. In our work sigmoid function was assigned to solve. It's have values from 0 to 1 in nonlinear profile function as given by equation (10).

$$sig(x) = \frac{1}{1+e^{-x}} \tag{10}$$

IOP Conf. Series: Materials Science and Engineering **717** (2020) 012011 doi:10.1088/1757-899X/717/1/012011

Three learning functions comprise the Scale conjugate method, Levenberg Marquardt algorithm and Resilient back propagation was comparable to selecting the highest accuracy of neural network model.

4. Experimental setup

The experimental setup is X-Y state actuator of ACAM machine as shown in Figure 4 for testing and simulation control system with the fault detection scheme. The major component consisting of lead screw with 5 mm pith diameter driven by DC servo motor 200 watt with incremental encoder 2500 PPR resolution and 24 Volts power supply. The linear encoder used as a feedback sensor to measure the position of the worktable.

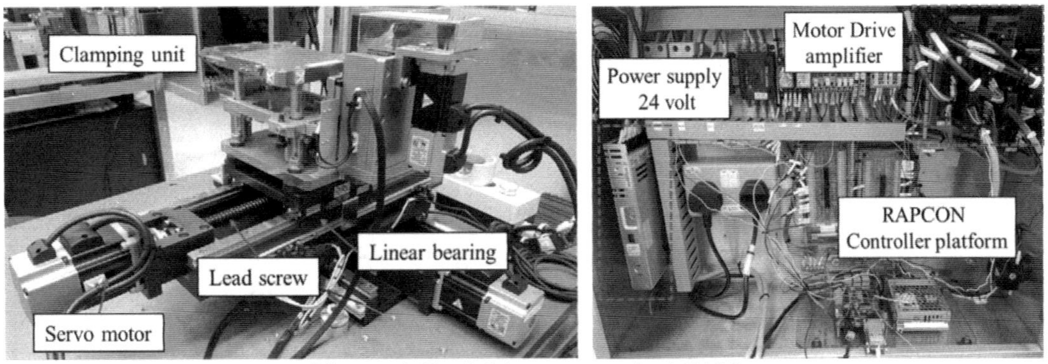

Figure 4. Feed driver system in ACAM

On each axis have supporting by four linear bearings. The motor driver amplifier will be received control signal from control unit before sent to the motor. In part of controller interface, RAPCON platform was applied with MATLAB/Simulink 2017a for design the propose method and used for collecting the data. In this paper, we have two encoders for validate the observer estimation result. For the linear bearing fault condition demonstrate on removing the lubricant left only 50% and starved lubricant condition compare with healthy bearing which are recommended for 2 ml to appropriate lubrication.

5. Results

5.1. Tracking response and observer performance

The controller, which designed as a PI servo system with state estimation using observer was examined by tracking response. For the position control of the worktable the design, test input has a reference at 50 mm, 58 mm and 66 mm, respectively that refer to the actual machine in the HGA process. The signal decomposition of ramp input 50 mm desire position and the other is step response. It was found that the system has an error in ramp input test, however, it reminds with an efficiency of position control which can be tracking reference input which can be compensated and eliminate error for the desire step response. On the other hand, the state observer is the ability for plant output estimation. Figure 5 is observation error and the state variable of the current signal, angular velocity, table velocity, angular position, and table position.

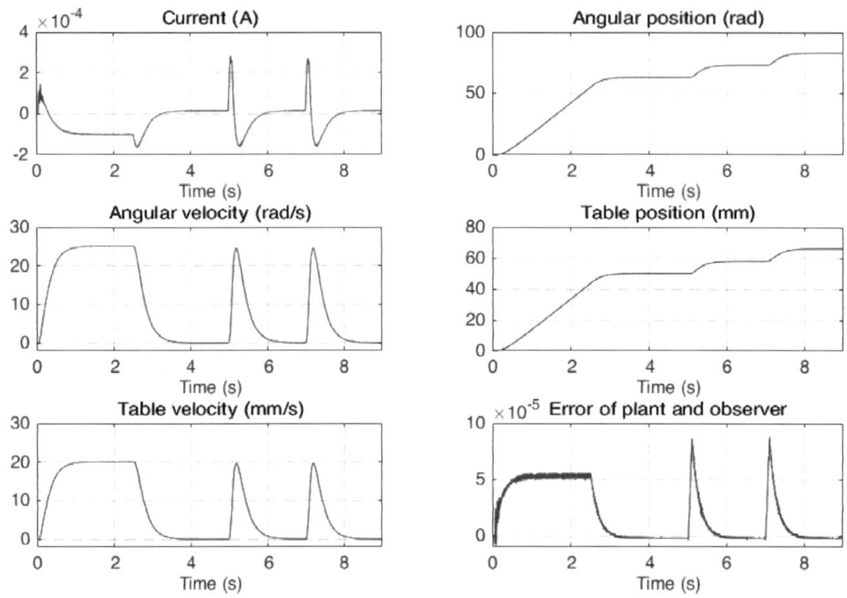

Figure 5. State variables estimate form the observer

5.2. Fault indication result on lubricant level

From the data manipulating and feature extraction into four statistical parameters the healthy condition, lubrication level 50% and starved lubricant of linear bearing were investigated in this research to perform the effect of linear bearing in the abnormal operating condition.

Table 1. ANN analysis

Algorithm	Epoch	Hidden layer No.	%Accuracy
Scale Conjugate	7	10	99.2
	7	15	99.2
	16	20	99.4
Levenberg Marquardt	20	10	99.4
	8	15	99.4
	6	20	99.7
Resilient Back propagation	9	10	99.2
	11	15	99.2
	10	20	99.2

The training process was compared to three learning algorithms with a varying hidden layer number that implement in MATLAB software. The results discussed in Table 1. The best performance of the ANN classifier can be measured on the model accuracy that is the error between the target and output must be the smallest. According to Table 1. Show clearly the ANN model which training by using the Levenberg Marquardt method offers the highest accuracy of 99.7% with a lower epoch. The confusion matrix illustrates the detail of class accuracy and correcting the prediction for the best ANN fault classification model as shown in Figure 6.

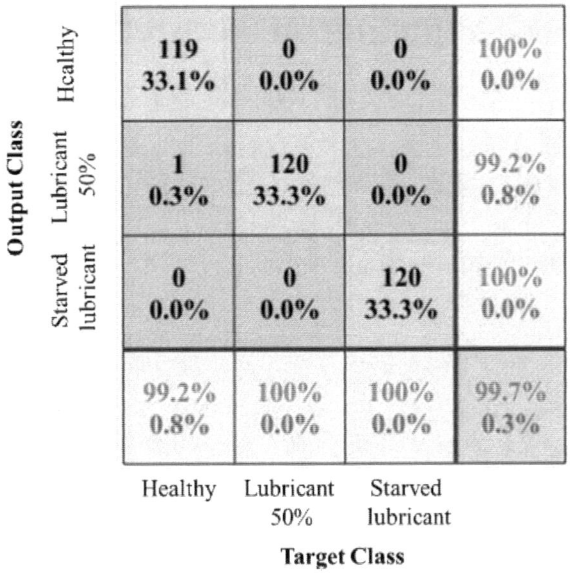

Figure 6. Validation confusion matrix of the best ANN model

6. Conclusion

This paper presents linear bearing fault detection for high-speed automation machine by integrating the intelligence method as an ANN with modern control system as PI servo system with an observer. The state variable includes current, angular velocity and table velocity was extracted to four statistical parameters for training input. The experimental results of position control system with an observer shows fast transient response, zero steady-state error and the state variable information from an observer. The ANN model using 20 hidden layers and Levenberg Marquardt algorithm can be identified three faults on the linear bearing which consisted of healthy condition, lubricant 50%, and starved lubricant with 99.7 % accuracy.

7. Reference

[1] Saengsri S, Prawanta S, Odngam S and Srisertpol J. 2017. *Proc. 2017 Int. Conf. on Circuits, Devices and Systems (Chengdu).* p6

[2] Aunsiri T, Numanoy N, Hemsuwan W and Srisertpol J. 2014. *Lecture Notes in Electrical Engineering.* **309** 921

[3] Dong X, Jian-qu Z and Feng W. 2012. *Procedia Engineering.* **29** 2853

[4] Jamaludin Z, Van Brussel H and Swevers J. 2009. IEEE Transactions on Industrial Electronics, 56(10) 3848.

[5] Dubey R and Agrawal D. 2015. *IET Science, Measurement & Technology*, **9(8)** 1016

[6] Zhao D, Wang T. and Chu F. 2019.*Computers in Industry.* **107** 59

[7] Chen Z, Deng S, Chen X, Li C, Sanchez R. V. and Qin H. 2017. *Microelectronics Reliability.* **75** 327

[8] Chommaungpuck P, Lawbootsa S and Srisertpol J. 2019. *WSEAS Transactions on Systems and Control.* **14** 31

Acknowledgments

This research was supported by Western Digital (Thailand) Co.Ltd and Suranaree University of Technology (SUT).

IOP Conf. Series: Materials Science and Engineering **717** (2020) 012012 doi:10.1088/1757-899X/717/1/012012

The Effect of Shaft Whirling on Accuracy of Rotating Coil Magnetic Measurement System

W. Thongtan, S. Odngam, T. Wanglomklang and J. Srisertpol

School of Mechanical Engineering, Suranaree University of Technology, Nakhon Ratchasima Province, Thailand

E-mail: jiraphon@sut.ac.th

Abstract. The electromagnetic in the storage ring is serve to focus the electron beam in vertical and horizontal axis along the beamline. Thus, precision of the electromagnetic field is a crucial step to considering for control the electron motion. The rotating coil is spread technique for measuring the electromagnetic field of magnet. However, the measurement uncertainties of the electromagnetic field are effect to controlled the electron motion before utilizing. This paper examines impact of the errors of the rotating coil magnetic measurement system. The whirling shape in each position along the length of the shaft is used for approximation uncertainty.

1. Introduction

The synchrotron light generator is used more electromagnetic in an electron storage ring. The inquiry on the condition of the electromagnetic field is important to consider. A rotating coil is an extensive approach to electromagnetic field measurement this method consists of the rotating coil speed control and voltage harmonic signal recording. The speed control of the rotating coil is effect to the nominal of the harmonic signal [1]. The PID controller is a conventional design technique, apply to speed control of the rotating coil [2]. The harmonic coil magnetic field measurement system and the measurement results for quadrupole magnets were detail [3]. In addition, the uncertainties of shaft diameter can be influenced by whirling of the shaft during rotation [4]. The characteristic of the harmonic signal which is measured from the search coil depends on the shaft whirling that can be investigated on misalignment in vertical and horizontal [5]. In order to minimize the whiling displacement considering on mass, length of shaft, radial and young's modulus is essential [6]. Moreover, natural frequency analysis and vibration, shape by using the modal test technique is important to determine the optimal speed [7]. The unbalance, mechanical misalignment and looseness phenomena can be detected by rotating coil technique [8]. This paper presents the verifying on whirling shaft imprecise of the electromagnetic field using a rotating coil approach. The experiment includes three topics. First topic demonstrates the natural frequency via the impact test, second topic is whirling shaft measurement that performs the motor speed at 200, 400 and 600 RPM and last topic simulates rotating coil magnetic measurement. The results of whirling shape can be used to approximate the measurement error that increasing the electromagnetic field measurement accuracy.

2. Equation of motion

The structure of the rotating coil is a coil that is packed in a cylindrical shaft, which is a circular shaft and supported by bearing, moves in a magnetic field with electric motor in a manner that revolves around itself at a constant speed. The mathematical model of shaft with bearing can be divided into two cases. The bearing is flexible object and a shaft is rigid object as first case, as shown in figure 1. The second case, considered bearing as a rigid object and a shaft as a flexible object, as shown in figure 2. where k_b is stiffness of the bearing, E is Modulus of Elasticity, L is the length of the shaft, M is the mass of shaft and I is the moment of inertia.

IOP Conf. Series: Materials Science and Engineering **717** (2020) 012012 doi:10.1088/1757-899X/717/1/012012

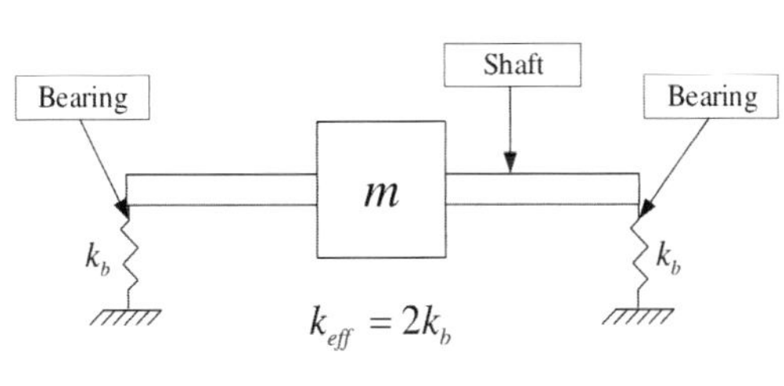

Figure 1. Rigid shaft on flexible bearing

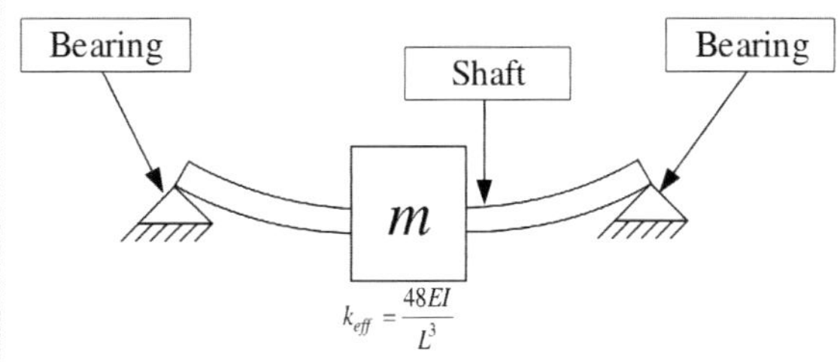

Figure 2. Flexible shaft on rigid bearing

The several whirling problems cause of shaft movement due to mass. Moreover, mass effect to unbalance while moving which unbalance influence to vibration force. The model of unbalance vibration is shown in figure 3. Unbalance of rotation, where k_{eff} is stiffness of shaft or bearing and f(t) is a force from unbalancing

$$f(t) = m\omega^2 e \sin \omega t \tag{1}$$

where m is the mass of shaft, e is centrifugation of the shaft from the center, ω is the angular velocity and *me* is the force that causes unbalance. Whirling of a shaft that caused by mass which the whirling causes the behaviour to move in various ways such as a circle or oval as shown in figure 4.

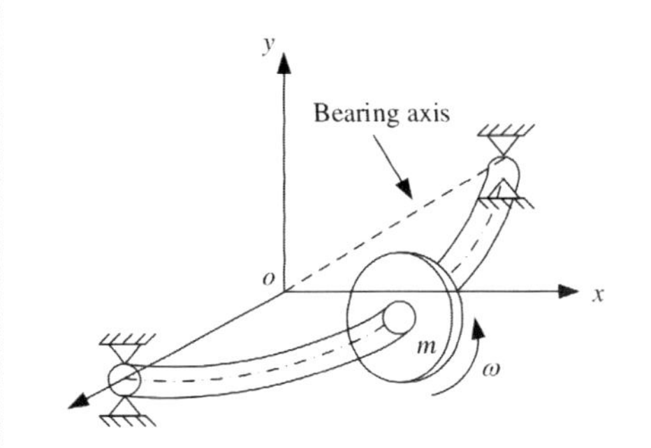

Figure 3. Whirling shaft mode

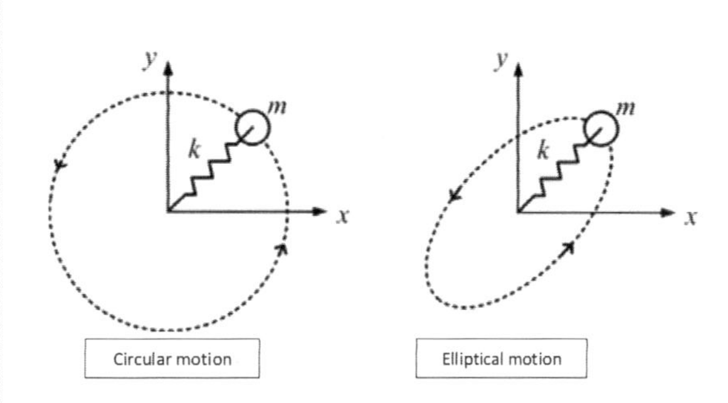

Figure 4. Shaft orbit.

The analysis of shaft includes of whirling shaft and shaft torsional in this case the conserving on 2 DOF and constant speed at 200, 400 and 600 rpm was investigated. The equation of motion in horizontal and vertical is shown in Eq. (2)-(3) respectively.

$$m\ddot{x} + c\dot{x} + kx = m\omega^2 e \cos \omega t \tag{2}$$

$$m\ddot{y} + c\dot{y} + ky = m\omega^2 e \sin \omega t \tag{3}$$

The equation of motion that can expose the natural frequency of shaft from free vibration without damping analysis from shown in Eq. (4)

$$\omega_{1,2} = (k/m)^{1/2} \tag{4}$$

The amplitude of centrifugal force due to shaft unbalance is $m\omega^2 e$ and the amplitude is dependent on rotating speed as shown in Eq. (5)-(6)

$$x(t) = X \cos(\omega t - \phi) \tag{5}$$

$$y(t) = Y \cos[(\omega t - (\phi + \tfrac{\pi}{2}))] = Y \sin(\omega t - \phi) \tag{6}$$

where X and Y are amplitude in horizontal and vertical, ω is rotating speed and ϕ is phase of displacement. The rotating coil magnetic measurement system has a symmetrical whirling of both axes through the magnetic field as shown in figure 5.

IOP Conf. Series: Materials Science and Engineering **717** (2020) 012012 doi:10.1088/1757-899X/717/1/012012

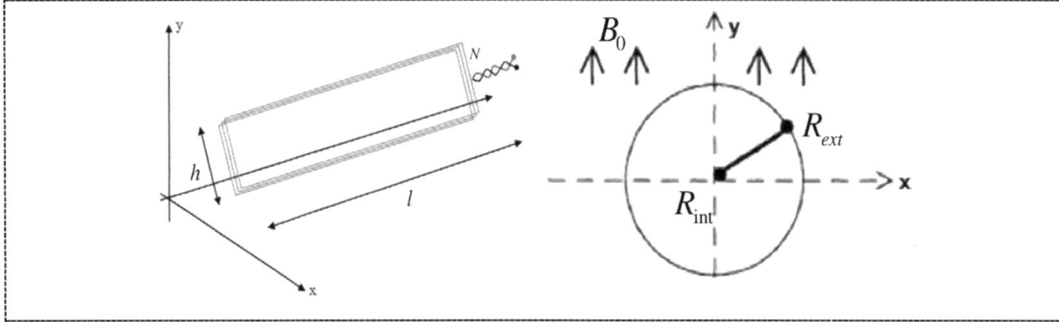

Figure 5.. Rotating coil in a dipole filed

The dipole magnetic flux can determine in Eq. (7).

$$\phi(\theta) = AB_o\, cos(\theta) \quad (7)$$

8he arithmetic of area of a coil as shown in Eq. (8).

$$A = NI(R_{ext} - R_{int}) \quad (8)$$

The electromotive force can determine from the rotating of the coil as shown in Eq. (9).

$$V = \mathrm{d}\phi/\mathrm{dt} = \omega AB_0\, cos(\theta) \quad (9)$$

Where B_0 is a magnetic field, N is a number of turns of the coil, I is the effective length of the coil and $R_{ext,}$ R_{int} is radial of rotation of the wires. From Eq. (9) The flux enclosed of rotating coil (ψ) to calculate as shown in Eq. (10)

$$\psi(\theta) = Nl \int_{R_{int}}^{R_{ext}} B_0\, cos(\theta) \quad (10)$$

where $\psi = volt \cdot sec = tesla \cdot m^2 = weber$

3. Experimental and Simulation results

The topic of results is divided into two sections: Section 1 describes the experimental setup for the investigation of natural frequency by impact testing and measuring the whirling of shaft horizontal and vertical direction and dynamic responses of the rotating coil and simulation results of impact test. Section 2 demonstrates the whirling of shaft measurement.

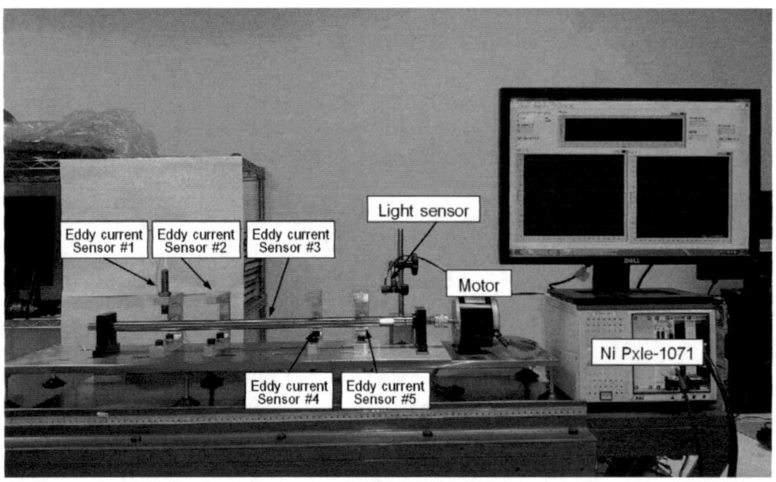

Figure 6. Experimental setup

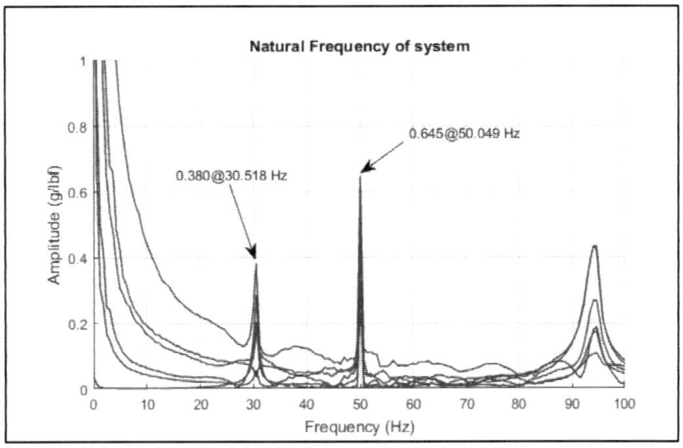

Figure 7. Frequency Response Function (FRF)

3.1. Impact test setup
The impact test section consists of the impact force hummer 1000 lbf sensitivity 1 mV/lbf with Dewetron DEWE-2601 and ± 50g sensitivity 105 mV/g accelerometer with FRF by Dewesoft X3.

3.2. Measurement whirling of shaft setup
The setup consists of NI PXIe-1071, the eddy current probe with an amplifier with 4 mm measurement range. The speed control system includes 24 Volt 250-watt DC motor, MATLAB/Simulink® and microcontroller board as shown in figure 6.

3.3. Impact test
The investigation of the natural frequency *used impact test by* roving sensor technique. The first mode of natural frequency is 1831.08 RPM (30.518 Hz) and the second mode is 3002.4 RPM (50.049 Hz) as shown in figure 7.

3.4. Measurement whirling of shaft
The speed of a rotating coil magnetic measurement system is good for a range of 200 – 600 rpm. duo to this speed is close to the actual speed that used in machine and far from the natural frequency. This section is measured at 200, 400 and 600 rpm in the vertical and horizontal direction and 5 measuring points in each speed as shown in figure 8. According to the experiment 1 to 3, it was found that the whirling shaft is a similarity. The amplitude in horizontal and vertical is increasingly dependent on rotating speed. The maximum amplitude of shaft whirling can be found at the 3rd measuring position and the first position shown the minimum amplitude in each speed. On the other hand, the bottom amplitude of shaft whirling in the vertical was indicate too high compared with the top. In addition, the amplitude at the third position illustrates in Table 1.

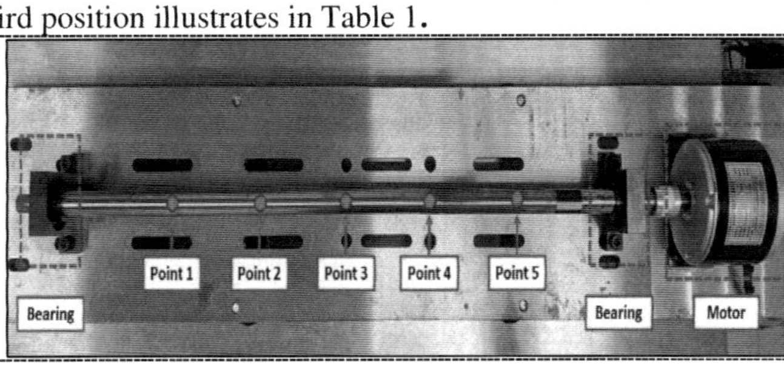

Figure 8. Measuring points

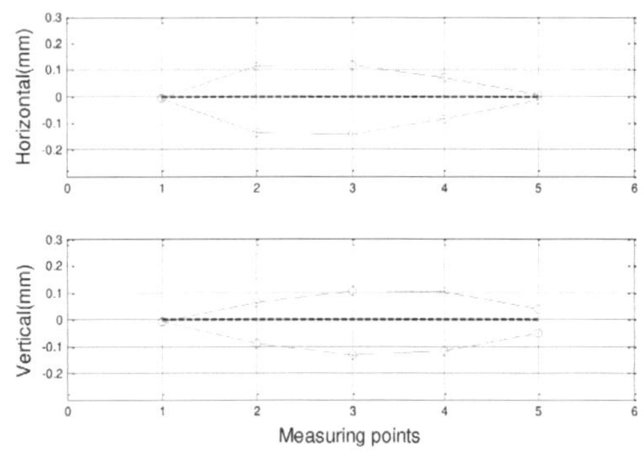

Figure 9. Whirling in horizontal and vertical of 600 RPM

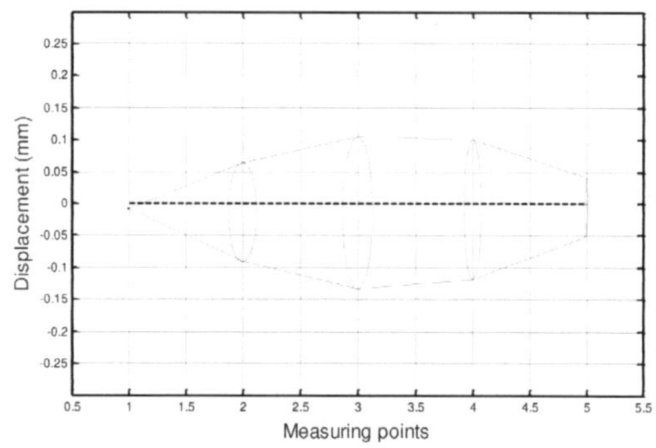

Figure 10. Whirling in radial of 600 RPM.

Table 1. The value of shaft whirling

Speed (RPM)	Horizontal(mm)		Vertical(mm)	
	Left	Right	Left	Right
200	-0.11648	0.13925	-0.19735	0.17456
400	-0.12049	0.14694	-0.14851	0.18653
600	-0.11632	0.14277	-0.15046	0.19058

3.5. Correlation of rotating coil

The experimental results 1 - 3 will be taken at the third measurement point because of the highest centrifugal distances as shown in Table 1 to simulate the shaft orbits and study the effects of shaft centrifuge on magnetic field measurements. The two polar magnetic field simulations with different magnetic field paths will be simulated.

Determined $\omega B_o = 1$ and area A has radial change according to the distance of whirling in the experiment through the magnetic field type Skew dipole (for observing the effect in horizontal field) and Normal dipole (for observing the effect in vertical field). The results are shown in figure 11.

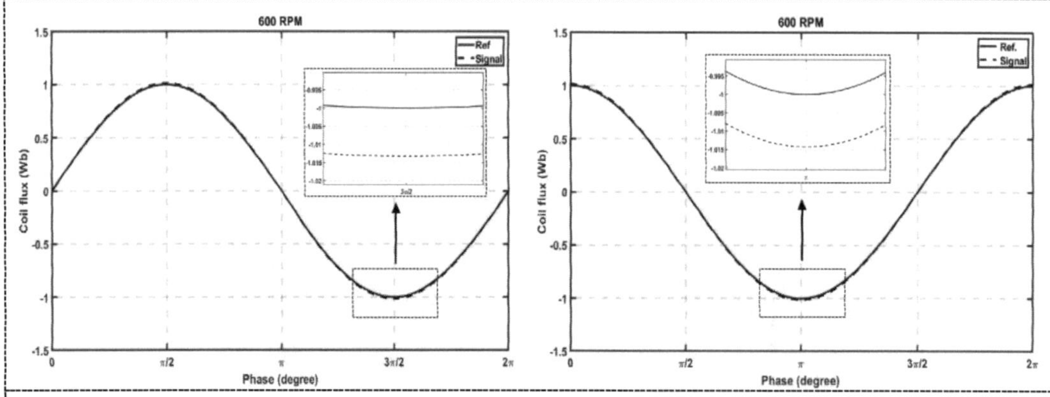

Figure 11. The harmonic of the magnetic flux when the rotating coil moving through skew dipole and normal dipole magnet at 600 RPM

The comparison between the simulation results of the ideal magnetic field measurement and the magnetic field measurement with whirling shaft at the rotational speed of 600 rpm in horizontal and vertical axis is shown in figure 11. The effect of the centrifugal force from the shaft weight. Causes the size of the radius (R_{ext}, R_{int}) to change. When measuring the magnetic field, therefore causing measurement discrepancies. The size of the error increase according to the shaft rotation speed as shown in Table 2.
Table 2, shown an error value of a magnetic field that measures in 2 axes. The simulation results show the error of vertical field is higher than the error of horizontal field at all speeds. Because the vertical axis has an earth's gravitational force acting on the shaft, therefore the vertical radial (R) is higher than the horizontal axis. The experiment found that if the accuracy of magnetic field measurement was needed, it is necessary to control the speed round of use. In order that rotating coil has minimum whirling.

Table 2. Percentage accuracy of the rotating coil

Speed (RPM)	**Error Horizontal** (%)	**Error Vertical** (%)
200	0.2399	0.3002
400	0.3151	0.3545
600	0.3153	0.3667

4. Conclusion
This paper presents effect of the whirling of the shaft that caused by weight, length and rotation speed round. For measuring of magnetic field intensity with rotating coil technique, measuring distance of the whirling of the shaft at horizontal axis and vertical axis at each position through the length of the shaft will can estimate the error of the measurement with this technique. In order to compensate the error that happen with the tool of the magnetic field intensity measurement for higher measurement accuracy.

5. Acknowledgment
The authors would like to express great thanks to Suranaree University of Technology (SUT) and Synchrotron Light Research Institute (Public Organization) for the research grant.

6. References
[1] Li L, Kang W, Li S, Zhang Z, Yin Bl, Zhou J and Shi C. 2014. *IEEE Transaction on Applied Superconductivity*. 24(3) 1
[2] Walckiers L. 2001. *Specialized Course on magnets*. The harmonic coil method. CERN Accelerator School.
[3] Li Li, Ying-Jie Li, Gan-Lin Ni, Wan Chen and Xian-Jing Sun. 2005. *Proc. the 2005 Particle Accelerator Conference (Tennessee)* 1844
[4] Purushotam K, Prabhakar K. P., Mithum K and Choudhary S. D. 2014. *Int. J. of Sci. & Eng. Res.* **5(11)** 1044
[5] Bottura, L. 2009. *CERN Accelerator School: Specialized Course on Magnets*. Field Measurements. CERN, Geneva, Switzerland.
[6] Zhan, H., Wang, M., Li, H., and Shi, X. 2016. *Proc. Int. Conf. on Computer Science & Education (Nagoya)* p 972
[7] Balasaheb K. T. 2014. *Int. J.of Sci., Eng. and Tech. Res..* **3(8)** 2091
[8] Zhou, H., Guo, X., Meng, T., Hong, X., and Yan, Y. 2013. *Proc. 2013 Fifth Conf. on Measuring Technology and Mechatronics Automation (Hong Kong)* p1087.

IOP Conf. Series: Materials Science and Engineering **717** (2020) 012013 doi:10.1088/1757-899X/717/1/012013

A 20.5 dBm Outphasing Class-F PA with Chireix Architecture at 3.5 GHz for RF Transmitter Front-end

Yusheng Sun and Sichun Du*

College of Computer Science and Electronic Engineering, Hunan University, Juzizhou Subdistrict, Changsha, China

*jt_dsc@hnu.edu.cn

Abstract. An outphasing class-F power amplifier (PA) with Chireix architecture is presented in Global Foundry (GF) 0.13μm Complementary Metal Oxide Semiconductor (CMOS) technology. The proposed circuit is composed of two branch class-F PA and Chireix power combiner with a floating load. The designed outphasing amplifier can provide 20.5 dBm output power from 1.8 V power supply at 3.5 GHz with 52.9 % power added efficiency (PAE). Wideband Code Division Multiple Access (W-CDMA) signal is implemented as input signal to simulate ACPR and the adjacent channel leakage ratios (ACLRs) at ±5 MHz are -20.5 dBc/-20.7dBc. After DPD (Digital Pre-Distortion) ACLRs can achieve -44.5 dBc/-44.7dBc at ±5 MHz. Furthermore, the chip area including testing pads is 1.98*1.62 mm^2.

1. Introduction

Modern wireless communication systems require high data rate,efficient spectrum utilization,necessitate low cost and power efficient solutions with high level of integration. There is a growing interest in integrating RF power amplifier (PA) in CMOS technology. Traditional linear PAs lose efficiency with power back-off for modulated signals with very high peak-to-average power ratio (PAPR). In order to solve these problem, single-branch and multi-branch PAs are proposed by researchers. Such as Envelope Elimination Restoration (EER), Envelope Tracking (ET) [1], Doherty [2] and Outphasing amplifiers [3]. Outphasing amplifier, as a PA with high efficiency and linearization and high-speed digital or analog signal processing, has become the key point of academic attention [4]. In this paper, a mult-branch class-F outphasing PA with passive non-ioslated combiner is proposed which improves efficiency at 3.5 GHz. The combiner is designed to expand the operating frequency of outphasing PA system. The compensating elements are selected in order to cancel the imaginary part of load modulation for 50 MHz operating frequency band centered around 3.5 GHz.

The paper is organized as follows, Section 2 introduces the outphasing theory and the Chireix combiner. In Section 3, we described the implementation of the Outphasing amplifier. The simulation result and analyzes are presented in Section 4, with conclusions following in Section 5.

2. Outphasing Theory and Chireix Combiner

The architecture of outphasing with a floating load-based Chireix combiner is demonstrated in Figure. 1. And Figure. 2 shows the conventional combiner and the combiner that we proposed.

2.1. The Traditional Outphasing Theory

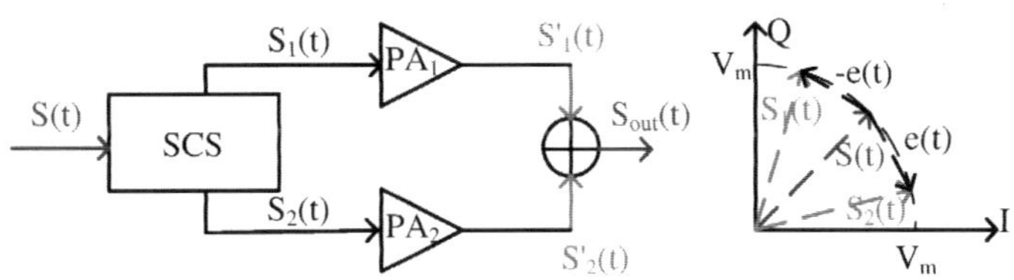

Figure 1: Traditional outphasing model and the signal vector decomposition.

Outphasing PA system [3], as shown in Figure. 1. The input signal can be expressed as:

$$S(t) = a(t)\, e^{j\theta(t)}, 0 \le a(t) \le V_m \tag{1}$$

This structure includes amplitude and phase modulation [i.e. $a(t)$ and $\theta(t)$], where V_m is the maximum amplitude of S(t), is decomposed [by a signal component separator (SCS)] into two constant-envelope phase-modulated signals:

$$S_1(t) = S(t) - e(t) = V_m e^{j[\theta(t) - \varphi(t)]} \tag{2}$$

$$S_2(t) = S(t) + e(t) = V_m e^{j[\theta(t) + \varphi(t)]} \tag{3}$$

in which

$$\varphi(t) = \cos^{-1}\left[\frac{a(t)}{V_m}\right] \tag{4}$$

$$e(t) = jS(t)\sqrt{\frac{V_m^2}{a^2(t)} - 1} \tag{5}$$

where $\varphi(t)$ is the dynamic outphasing angle. The two signals $S_1(t)$ and $S_2(t)$ are separately amplified.

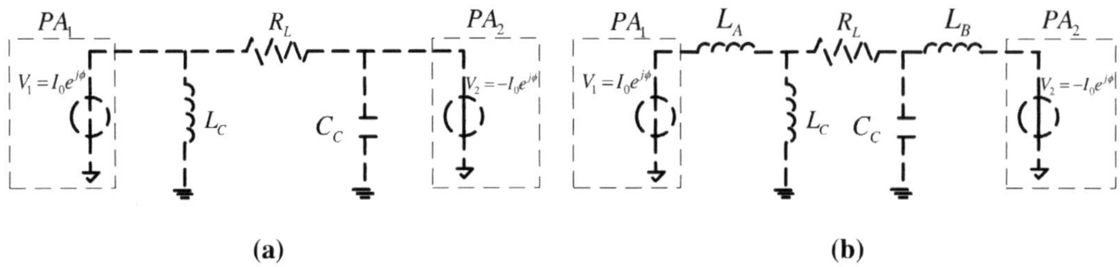

Figure 2: (a) With single compensation components, (b) With double compensation of components.

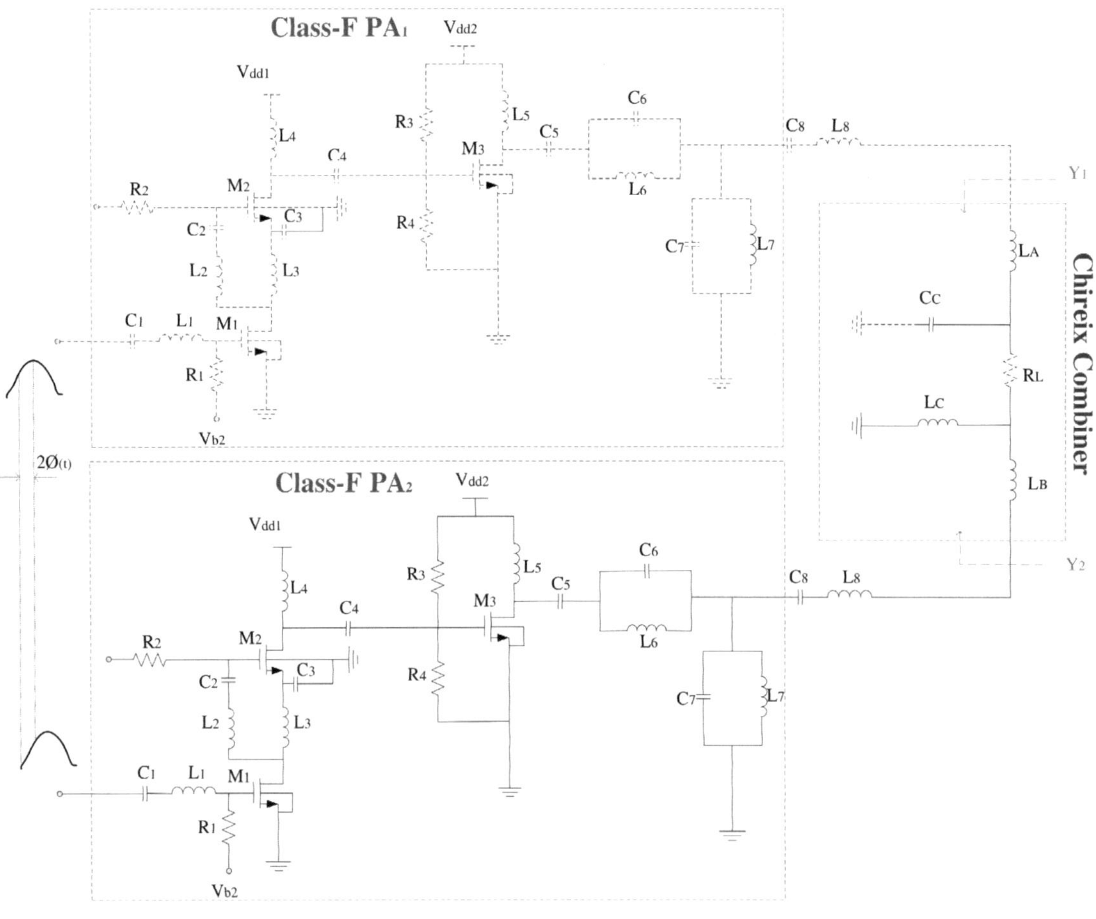

Figure 3: Circuit schematic of the proposed outphasing class-F PA with chireix structure.

2.2. The Design of Chireix Combiner

Figure. 2 (a) illustrates the conventional chireix combiner structures with single compensation components. All of these structures use inductor Lc and capacitor Cc to compensate for phase-varying reactance when the two branches are different in phase. Therefore, PAs can acquire maximum efficiency when the outphasing angle ϕ of two branches signal are offset each other. And the values of inductance and capacitance can be derived as [5]

$$C_c = \frac{sin2\phi}{R_L\omega_0}, L_c = \frac{R_L}{\omega_0 sin2\phi} \tag{6}$$

However, the angle ϕ_1 from PA_1 and ϕ_2 from PA_2 do not eliminate with each other in actual circuit test. To offset excess phase angle,double compensation of components structure is proposed in Figure. 2 (b).

3. Circuit Implementation

Figure. 3 depicts the circuit schematic of the proposed outphasing amplifier using the LC compensation Chireix combiner. Notice that two stage class-F PA provide different outphasing angle to achieve high efficiency and bandwith at both peak and back-off power levels [6]. In addition, C_1, L_1 and C_8, C_8 are the input and output resonant network. And L_A, L_B are added into Chireix combiner to achieve outphasing angle offset each other of two branch of the PA.

4. Simulation Results and Analysis

In order to facilitate the simulation, this work plans to use PC software of MATLAB to generate digital W-CDMA data stream and then gets through DAC module to produce I/Q signals as system base-band signals. Figure. 4 shows the block diagram of simulation processing scheme. The output power and power gain with respect to input power are described in Figure. 5 for

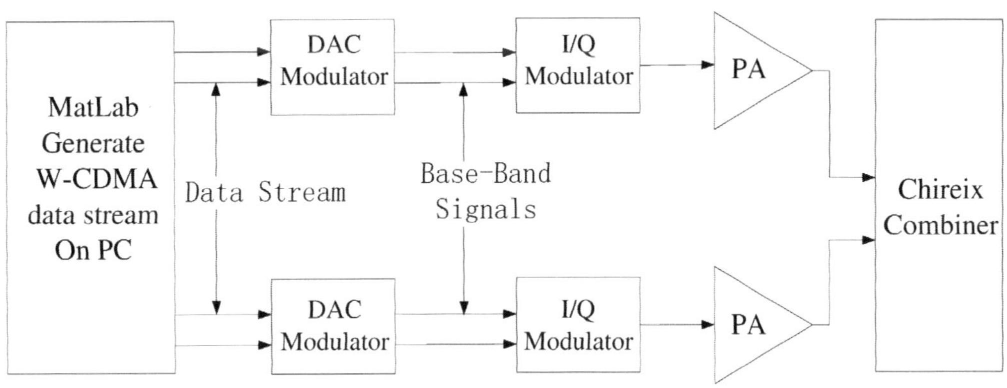

Figure 4: Block of diagram of simulation processing scheme.

proposed outphasing amplifier. This figure manifests that the outphasing amplifier achieves 20.5 dBm peak output power with a 1.8 V power supply. Meanwhile, the 34 dB of power gain is presented when the input power is below -15 dBm. Since the proposed PA is based on the structure of the combiner, the input matching parameter S_{11} and S_{11} of the amplifier is considered as shown in Figure. 6. It is clear that $S_{11} = -6.5$ dB and $S_{22} = -11.8$ dB

IOP Conf. Series: Materials Science and Engineering **717** (2020) 012013 doi:10.1088/1757-899X/717/1/012013

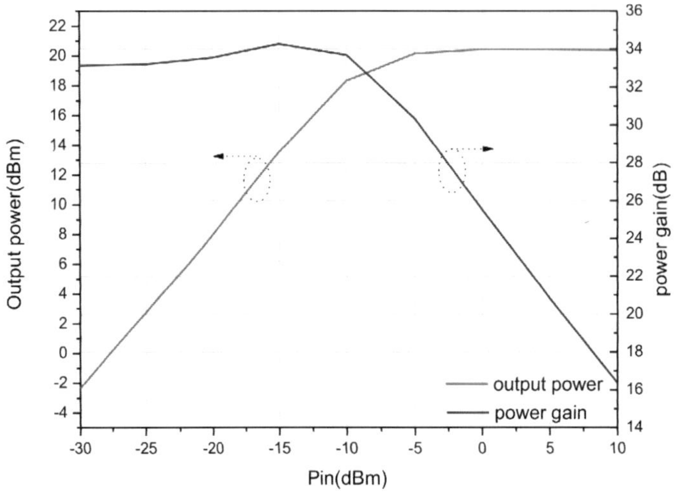

Figure 5: Output powers and power gains with respect to input powers.

at 3.5 GHz frequency point. Consequently the input and output matching also realizes better matching results. Figure. 7 shows that the stability factor always $K > 1$, that is, the PA is absolutely stable at 3.5 GHz frequency point. Figure. 8 illustrates that the single side branch PA achieves peak PAE of 52% and maximum outphasing PAE of 52.9% with peak output power of 20.5dBm with chireix combiner. Furthermore,the output spectrum for PA is provided in Figure. 9. Experimental verification is performed by two independent in-phase/quadrature (I/Q) modulating signal which are generated by MATLAB software. In this simulation experiment, pre-distortion and non-pre-distortion linearization technique are carried out respectively on PA circuit system.

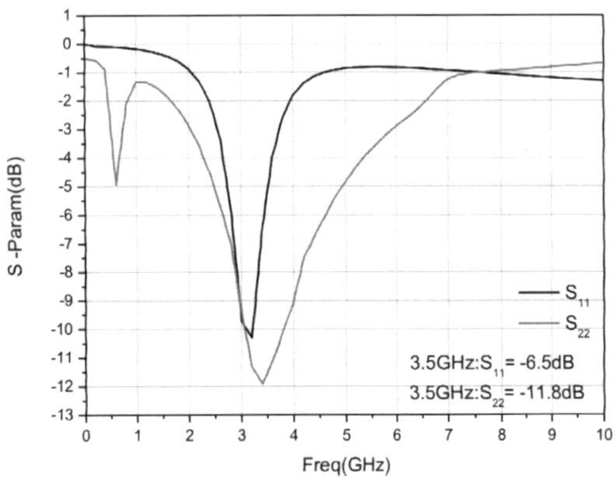

Figure 6: S parameters of the proposed power amplifier.

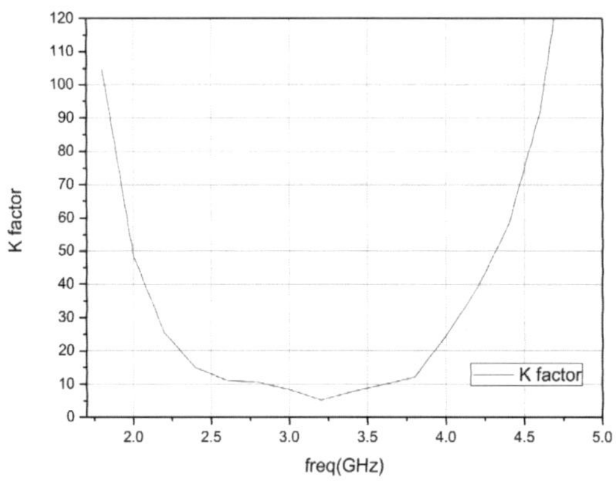

Figure 7: K factors of the proposed power amplifier.

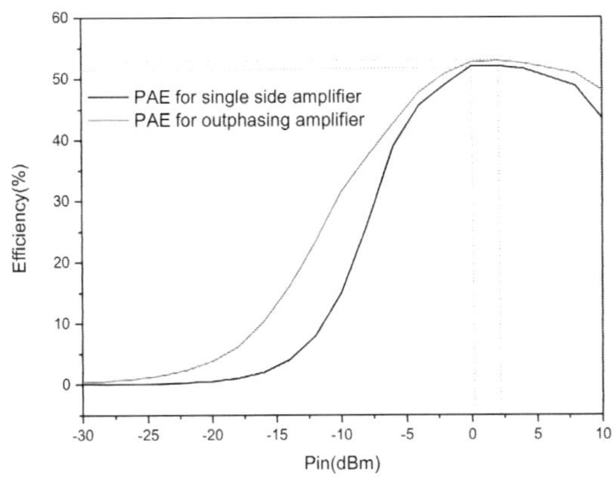

Figure 8: Power added efficiency for single side and outphasing power amplifier.

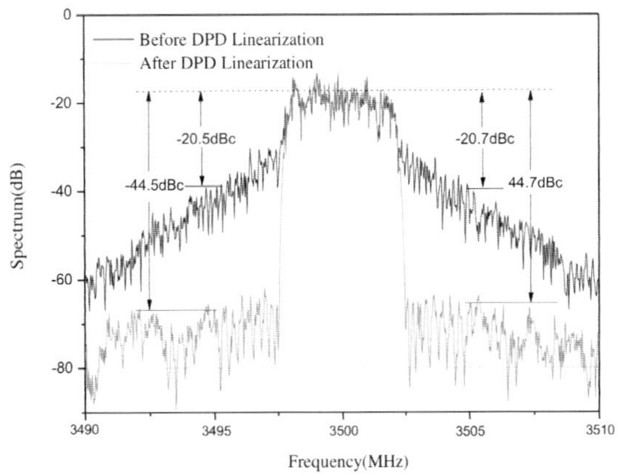

Figure 9: PA output spectrum at 3.5GHz with and without DPD technique.

The designed outphasing PA is simulated with 5 MHz W-CDMA signal. Simulated lower/upper adjacent channel power ratio levels are -20.5 dBc/-20.7 dBc at 5 MHz offset from the carrier center frequency. After DPD technique of linearization, lower/upper adjacent channel power ratio levels are -44.5 dBc/-44.7 dBc and the ACPR improvement of 20/20 dB. These result indicate that the system has good linearization and less adjacent channel leakage. In the end, the layout of proposed outphasing amplifier is presented in Figure. 10, which occupies a chip size of 1.98*1.62 mm^2. Table 1 makes a comparison with published outphasing PA, including various signal modulation, as well as the performance of different processes. Because of the compromise, this work is still very advantageous which is based on the summary of the Table 1.

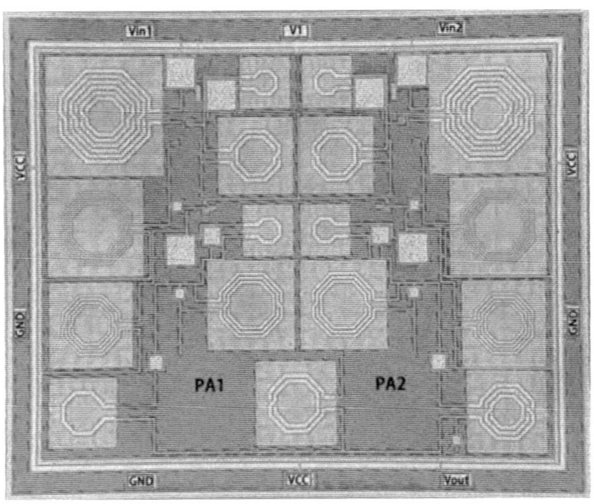

Figure 10: The layout of the proposed outphasing class-F PA with Chireix structure. The total chip size is 1.98*1.62 mm^2

Table 1: Performance comparison

Ref	Pout (dBm)	PAE (%)	V_{DD} (V)	Gain (dB)	Die Size (mm^2)	f_0 (GHz)	ACPR (dBc)	Modulated signal	Technology
[7]	20	42	1.8	N/A	N/A	0.81	-45/-45	CDMA	0.18μm
[8]	25.2	40	2.5	N/A	1.08	0.8-2	-46/-46	WCDMA	0.13μm
[9]	27.7	45	2.5	N/A	4	2.4	N/A	OFDM	65nm
[10]	29.5	46.7	2.4	N/A	N/A	2.4	-51.08/-50.4	LTE	45nm
[11]	20	56	2.5	N/A	1.72	0.9	-50/-50	WCDMA	90nm
Ours	20.5	52.9	1.8	34	3.2	3.5	-44.5/-44.7	WCDMA	0.13μm

5. Conclusion

This paper presents a 0.13μm class-F outphasing PA with a non-isolated combiner, the outphasing PA based on lumped components and that also helps to improve the efficiency at back-off area. This structure can effectively overcome the problem of efficiency degradation when power is backed up. And this task achieves a peak power of 20.5 dBm, 52.9 % PAE. By comparison, the results have more advantages than other equivalent circuits and indicate the potential of the approach. Although further effort is required to linearize the input and output and PAE of the amplifier. Circuits design for improved linearity and PAE will be focused in the future paper.

6. References

[1] Cripps S C 2000 RF Power Amplifiers for Wireless Communications *IEEE Microw. Mag.* **1** 64

[2] Doherty W H 2014 A new high efficiency power amplifier for modulated waves *Proceedings of the Institute of Radio Engineers* **24** 1163

[3] Chireix H 1935 High Power Outphasing Modulation *Proceedings of the Institute of Radio Engineers* **23** 1370

[4] Van der Heijden M P, Acar M, Vromans J S and Calvillo-Cortes D A 2011 A 19 W high-efficiency wideband CMOS-GaN class-E Chireix RF outphasing power amplifier *IEEE MTT-S International Microwave Symposium* (Baltimore: IEEE)

[5] Lee S, Nam S 2010 A CMOS Outphasing Power Amplifier with Integrated Single-Ended Chireix Combiner *IEEE Trans. Circuits Syst. II-Express Briefs* **57** 411

[6] Calvillo-Cortes D A, et al 2013 A Package-Integrated Chireix Outphasing RF Switch-Mode High-Power Amplifier *IEEE Trans. Microw. Theory Tech.* **61** 3721

[7] Hung T P, Choi D K, and Larson L E, et al 2007 CMOS Outphasing Class-D Amplifier With Chireix Combiner *IEEE Microw. Wirel. Compon. Lett.* **17** 619

[8] Presti C D, Carrara F 2009 A 25 dBm Digitally Modulated CMOS Power Amplifier for WCDMA/EDGE/OFDM With Adaptive Digital Predistortion and Efficient Power Control *IEEE J. Solid-State Circuit* **44** 1883

[9] Godoy P A, Chung S W, Barton T W, et al 2012 A 2.4-GHz, 27-dBm Asymmetric Multilevel Outphasing Power Amplifier in 65-nm CMOS *IEEE J. Solid-State Circuit* **47** 2372

[10] Banerjee A, Hezar R, Ding L, et al 2014 A 29.5 dBm class-E outphasing RF power amplifier with performance enhancement circuits in 45nm CMOS *European Solid State Circuits Conference* (Venice: IEEE)

[11] Moloudi S, Abidi A A 2013 The Outphasing RF Power Amplifier: A Comprehensive Analysis and a Class-B CMOS Realization *IEEE J. Solid-State Circuit* **1** 64

Acknowledgment

This work was supported in part by the National Natural Science Foundation of China (No.61674054) with the Natural Science Foundation of Hunan Province, China (No.2017JJ2049) and was funded by the Innovation platform fund project of Education Department of Hunan Provincial (No.17K021).

IOP Conf. Series: Materials Science and Engineering **717** (2020) 012014 doi:10.1088/1757-899X/717/1/012014

Study on Simulation of Warning Method and Evaluation of Car-following for Intelligent Drive

Ren Nv-er, Jiang Rong and Cheng Xu

China Automotive Technology and Research Center Co., Ltd. 68, East Xianfeng Road, Dongli District, Tianjin, China

jiangrong@adcsoft.cn

Abstract. Car-following behavior is one of the main driving behaviors of drivers in road traffic environment, the simulation experiment for car-following early warning of intelligent driving can provide driver with safe and efficient verification before Real Vehicle Experiments. The focus of the research is how to detect car-following distance by LiDAR and evaluate the car-following warning method under the simulation environment. In this paper, the mathematical model of LiDAR was established in the simulation environment so that we can obtain the Time Headway changing to warn driver to accelerate or not in real time. Next in importance, in Unreal Engine 4 virtual scene, by using the special collision detection box to simultaneously detect the longitudinal and lateral relative positions of the ego vehicle and the front vehicle, we can complete the comprehensive evaluation of car-following behavior in both horizontal and vertical directions. Ultimately, to verify the validity and reliability of this method, hardware-in-the-loop simulation of car-following warning method was realized by connecting the driving simulator. As a consequence, by using the method of this paper, we provide a feasible technical scheme for intelligent driving vehicles in respect of multi-lane car-following warning.

1. Introduction

In recent years, with the rapid development of intelligent driving, the active safety technology of automobile is the core part of intelligent driving automobile. Since car-following behavior is one of the main driving behaviors of drivers in road traffic environment [1], it is very necessary to effectively identify in the current lane, left and right lanes while following cars. Appropriate car-following early warning method can reduce traffic accidents to some extent. At present, there are four principal methods for obstacle avoidance technology, which includes infrared sensor, ultrasonic sensor, laser sensor and visual sensor. For instance, Darmstadt University used Camera and millimeter wave radars successively during the period from 2011 to 2016, and the method of fusing the two sensor measurements was used in it's latest published article [2-5]. To get along with, Chang'an University and Central South University also made full use of the precise characteristics of radar ranging to get the relative position of obstacles at home. The difference two universities adopt two different kinds of radar, millimeter wave radar [6] and LiDAR [7], respectively. As a safe, efficient and economical method [8], simulation has been paid more and more attention in the verification of various intelligent driving control algorithms [9]. 3D simulation scenario and mathematical model of LIDAR has been established for intelligent driving under UE4(Unreal Engine 4). Therefore, the distance between ego vehicle and preceding vehicle in-lane, left lane preceding vehicle and right lane preceding vehicle could be detected. It is effective to evaluate car-following behavior by comparing the detected distance with the threshold of the car-following algorithm [10].

2. Multi-lane vehicle following model[10]

The role of driver-following model is to simulate the vehicle control characteristics by the driver in the process of car-following. Numerous studies have shown that the two parameters, Time Headway(THW) and inverse of Time to Collision, are Control Objectives for driving process, especially stable car-following process. Since a lot of research has been done on the frontal collision warning algorithm both at home and abroad [10-16], the multi-lane car-following model in reference [10] was adopted as shown in Formula (1).

$$\mathrm{RP(k)} = \frac{a_{des}(k)}{v_{ego}(k)} = K_d \cdot \left[\overline{THW_m}(k) - THW_d\right] + K_v \cdot \overline{THW_m}(k) \cdot \overline{TTCi_m}(k) \quad (1)$$

Where, k is the sampling sequence number of the data. RP, namely risk perception index, is risk perception index defined by forward collision warning algorithm. RP means the ratio of the driver's expected acceleration to the vehicle's speed. Appropriate RP value could be set as the alarm threshold of the forward collision warning system. THW_d is expected Time Headway of driver, K_d and K_v are dynamic parameters relating to the driver's characteristics, $THW_d = 2.16s$, $K_d =- 0.0012s^{-2}$, $K_v = 0.065$. And $\overline{THW_m}(k)$, $\overline{TTCi_m}(k)$ was calculated by formula (2) and formula (3) respectively.

$$\overline{THW_m}(k) = \sum_{i=1}^{n} \rho_i(k) \cdot THW_m(k - n + i) \quad (2)$$

$$\overline{TTCi_m}(k) = \sum_{i=1}^{n} \rho_i(k) \cdot TTCi_m(k - n + i) \quad (3)$$

$$THW_m = \frac{d_m}{v_{ego}} \quad (4)$$

$$TTCi_m = \frac{v_{rel-m}}{d_m} \quad (5)$$

$$v_{rel-m} = \lambda_l \cdot v_{rel-l} + \lambda_h \cdot v_{rel-h} + \lambda_l \cdot v_{rel-r} \quad (6)$$

$$d_m = \lambda_l \cdot d_{y-l} + \lambda_h \cdot d_{y-h} + \lambda_l \cdot d_{y-r} \quad (7)$$

$$\sum_{i=1}^{n} \rho_i(k) = 1 \quad (8)$$

Where, v_{rel-m} is synthetic relative velocity in condition of Multi-lane car following which could be calculated by Formula (6), d_m in formula (7) is the longitudinal relative distance. v_{rel-m} and d_m are calculated by the relative velocity and relative distance of the nearest target in each lane according to the comprehensive weight of the target. v_{ego} is the speed of the ego vehicle. Weighted-value matrix $\rho(k) = [0.05\ \ 0.1\ \ 0.15\ \ 0.25\ \ 0.45]$ was adopted to simplify the formula (8).

3. Modeling of LiDAR

Before establishing the mathematical model of LiDAR, we should simplify laser beam of the transmitting at first step, then modeling LiDAR by describing main functional parameters. And we modeled LiDAR which is currently used in automatic driving vehicles based on Time of Flight method.

3.1 Prerequisites for modeling

Single beam of LiDAR is an approximate rectangular facula with a certain cross-sectional area, and the laser facula area increases as distance from ego the vehicle increases. The maximum edge length of the facula presented at the maximum detection range of the LiDAR is about cm level. Accordingly, compared to common objects in Driving Environment, the beam of laser radar can be seen as a line without thickness. For this reason, the laser radar beam is abstracted as a ray model which could facilitate the modeling of LiDAR function parameters.

3.2 Modeling of LiDAR function parameters

3.2.1 Detection range and resolution model. LiDAR sensors can identify the details of a few centimeters at more than 100 meters, and its detection range is related to object reflectivity and transmission medium. Therefore, the detection range of LiDAR is the superposition of theoretical detection value and distance error.

The theoretical detection value of LiDAR in this paper satisfies the following Formula (9)

$$R_{truth} = \begin{cases} 0, \text{ No object was detected or less than the minimum detection range} \\ R_t, \text{ minimum detectable range} \leq \text{distance} \leq \text{maxmum detectable range} \end{cases} \quad (9)$$

Here, the detection value of LiDAR is set to 0 under the condition of invalid detection range of LiDAR. When distance between measured object and LiDAR is within the range of effective detection range, that is, between the minimum and maximum detection distance, the normal distribution model is used to simulate the detection capability of LiDAR, as shown in the following formula. (10)

$$f(R_{Msr}) = \frac{1}{\sqrt{2\pi}\sigma} e^{-\frac{(R_{Msr}-R_{truth})^2}{2\sigma^2}} \quad (10)$$

Where, R_{truth} is the theoretical value detected by LiDAR while R_{msr} is the practical value, and the units of the above two variables are cm; The standard deviation σ represents the ranging precision in cm. That is to say, the measurement value, theoretical detection value and range resolution of LiDAR satisfy the normal distribution relationship. We can't solve the equation directly because it belongs to transcendental equation. Therefore, Box-Muller method was used to approximate the solution of the unknown variable R_{msr}, which can be realized in the simulation software.

3.2.2 Models of horizontal field of view angle, horizontal field of view angular resolution and rotation frequency. Horizontal field of view angle, horizontal field of view angular resolution and rotation frequency of LiDAR are correlated to some extent. As the mechanical LiDAR adopts the cycle scanning mode, horizontal field of view angle range of LiDAR is 360 degrees. The model used in this paper is as formula. (11)

$$0° \leq \theta_h < 360° \quad (11)$$

θ_h is the horizontal field of view angle in degree.

The rotation frequency is a discrete integer that varies in a certain range, and its relationship with horizontal resolution is described as the following formula (12)

$$\sigma_h = \frac{N}{C_h} \quad (12)$$

In the formula, σ_h is a horizontal resolution in units of degree; N is the initial rotation frequency in Hz; C_h is a constant connecting horizontal resolution and rotation frequency, which is related to the number of transmitting points per second in Hz/degree.

3.2.3 Models of vertical field of view angle, vertical field of view angular resolution and channel number. Multi-channel laser radar usually contains 16 channels, 32 channels, 64 channels and 40 channels, which are finite discrete values. So the number of channels in this paper is an enumeration type.

For the existing LiDAR products, the maximum and minimum field of view angles are not all symmetrical. The model used in this paper is shown as formula (13)

$$\theta_{v\_min} \leq \theta_v \leq \theta_{v\_max} \quad (13)$$

In fact, the vertical field of view resolution of LiDAR with 16 channels is uniform, and that of LiDAR with more than 16 channels is discrete and non-uniform. The vertical field of view angle model is described as formula (14)

$$\sigma_v = (\theta_{v1}, \theta_{v2}, \ldots \theta_{vn}) \quad (14)$$

Among them, σ_v is the set of vertical field of view angular resolution, θ_{v1} is the minimum field of view angle, θ_{v2} is the sub-minimum field of view angle and θ_{vn} is the maximum field of view angle. n indicates the channel number of LiDAR, dimensionless.

The above models were embedded in the software and initialized with functional parameters. The instantiated LiDAR model was applied to acquire data in simulation scenario.

4. Construction of simulation environment

In this paper, according to the intelligent driving and car-following scheme, a multi-lane car-following model was selected, and a mathematical model of LiDAR has been established in virtual environment to detect the car-following distance in real time. In order to verify the effectiveness of the proposed method, Unreal Engine was used to accomplish the simulation and evaluation of multi-lane car-following.

4.1. Modeling of 3D models in intelligent driving simulation environment

Under UE4 Development Environment, typical driving scenarios are built based on real traffic scenarios by utilizing function modules such as UE4 blueprint editor, AI, behavior tree and navigation. Switching of traffic signals and AI Logic in vehicles except ego vehicle were added there. The start-stop, driving, following and lane-changing driving behavior of vehicles that obey the traffic rules were simulated with the blueprints visual script. The LODs (Levels of Detail) algorithm has been adopted for fast rendering complex scenes and optimizing resources.

4.2. Generation of LiDAR point cloud data

High accuracy point cloud data can be obtained by using LiDAR model in the second part. However, the basic research on the organization, management and visualization of point cloud data has always been an important research topic in the field of laser scanning technology since it is an important basis for follow-up analysis and treatment. Real-time vehicle distance detection, which could provide reference data for driver behavior decision-making at intelligent driving, was accomplished by LiDAR model.

In the process of simulation experiment, it requires higher read and write access speed when we want to obtain the detected point cloud data in real time. Shared memory is the fastest form of inter-process communication. Compared with reading file contents from the disk or database in speed of execution, shared memory has an absolute advantage. The operating system maps a memory segment in the address space of several processes, so that several processes can read and write large data quickly in that memory segment without calling operating system functions. It greatly improves the communication speed. Figure 1 illustrates how can we obtain the simulation point cloud data in real time.

4.3. Overall design of simulation scheme

The real-time speed of ego vehicle $v_{ego}(k)$ has been collected at 60 frames per second. In addition, the corrected $\overline{THW_m}(k)$ and $\overline{TTCi_m}(k)$ on multi-lane road were captured by mathematical model of LiDAR. RP(k) was calculated by substituting $\overline{THW_m}(k)$ and $\overline{TTCi_m}(k)$ into the car-following model. If RP(k) is less than 0, deceleration signal will be sent to driver in simulation environment. on the contrary, acceleration signal will be sent. Figure 2 is a flow chart for driving decision-making.

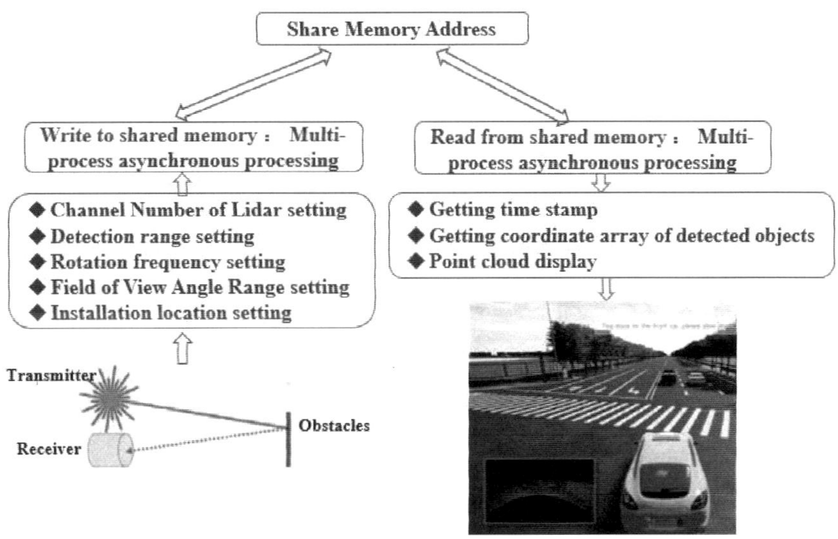

Figure 1. Obtain point cloud data by simulation

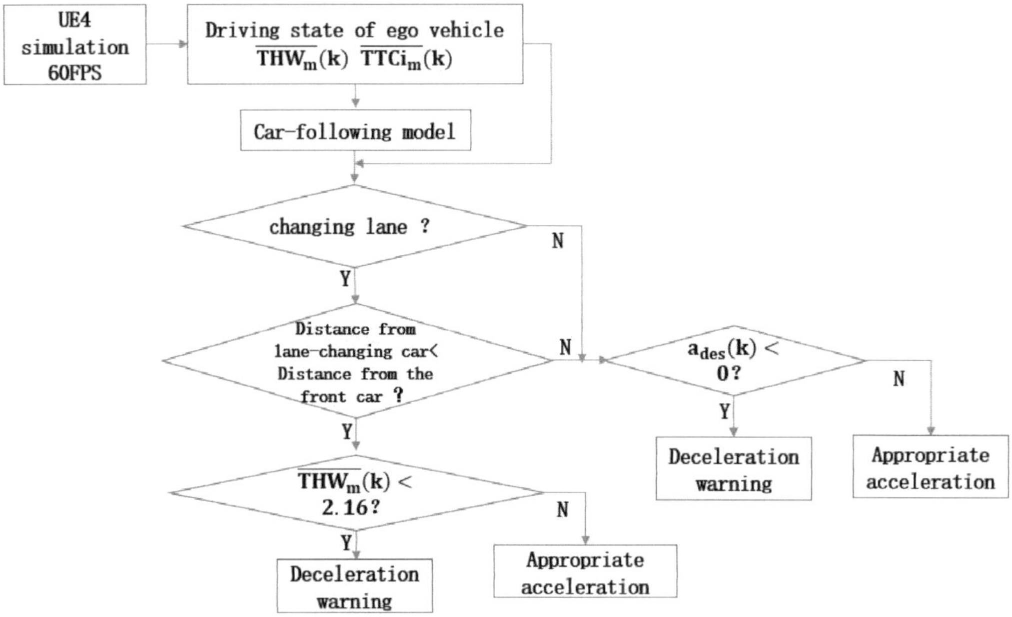

Figure 2. Flow chart for driving decision-making

4.4. Hardware configuration of driving simulator

The hardware of driving simulator adopts Logitech G29 force feedback steering wheel pedal suit which includes steering wheel, a manual gear, and a racing pedal with clutch. The steering wheel has 900 degrees steering range, which can simulate real driving behavior to the greatest extent, and also provides programmable keys and direction control keys. Evaluation of Car-following simulation method.

The car-following model used in the preceding paper is mainly aimed at longitudinal direction. In order to evaluate car-following behavior more comprehensively, the longitudinal and lateral relative positions of the ego vehicle and the front vehicle can be judged simultaneously by generating collision detection boxes in the virtual engine. In the simulation scenario, a row of collision detection boxes were generated in the rear of the followed vehicle according to the trajectory and the width of the vehicle body in front. It's convenient that the volume, number, generation interval and disappearance time of collision detection boxes can be customized according to the fine degree of evaluation. Therefore, number of detection boxes collided by ego vehicle showed clearly whether car-following behavior is good or bad.

In the hardware-in-the-loop simulation environment, comparing the $\overline{THW_m}(k)$ captured by the mathematical model of LiDAR with THW_d all the time, we could remind the driver to control the speed in real time. Figure 3is the output data of an optional car-following simulation. It is not difficult to see from the Figure 3that at the beginning of the simulation experiment, the Time Headway $\overline{THW_m}(k)$ was much longer than the expected Time Headway THW_d. The UI interface will prompt the driver to "accelerate properly", as shown in Figure 4. The $\overline{THW_m}(k)$ will decrease with the acceleration of the driver, even less than THW_d. The UI interface will prompt the driver to "slow down", as shown in Figure 5. Due to the continuous evaluation of driver's driving behavior by simulation system, drivers will gradually enter the state of stable car-following.

At the same time, the number of detection boxes collided by the ego vehicle was detected and compared with the total number of detection boxes generated online, the results are shown in Figure 6. From Figure 6, we can see that the number of detection boxes collided by ego vehicle is zero because the $\overline{THW_m}(k)$ is much greater than the THW_d. With the increase of time, the proportion of detection boxes collided gradually increases, and finally tends to be stable. This also shows the stable car-following situation of Figure 3from the side.

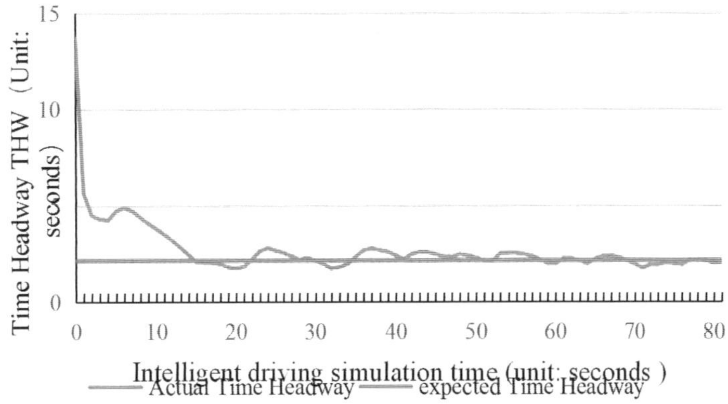

Figure 3. Real-time Time Headway of Intelligent Driving Simulation

Figure 4. $\overline{THW_m}(k)$ was much longer than the THW_d

Figure 5. $\overline{THW_m}(k)$ was less than the THW_d

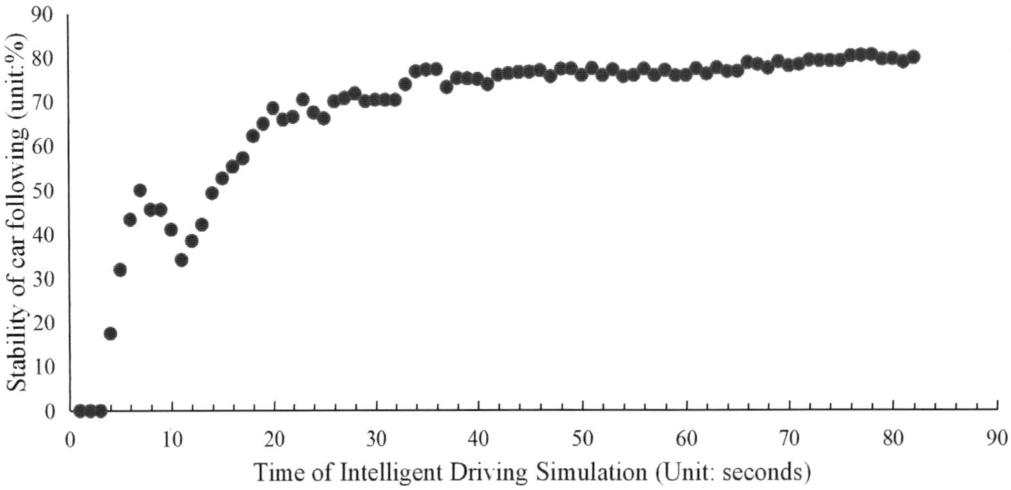

Figure.6 Evaluation of Car-following Simulation Method

5. Conclusion

With the continuous improvement of artificial intelligence, intelligent driving has attracted more and more attention. However, the complex changes in the traffic system make the safety of intelligent driving face challenges. Aiming at the deficiency of existing research, based on the mathematic model and car-following model of LiDAR, the following work is carried out in this paper: Under the simulation environment of UE4, the vehicle in front is detected by using the mathematical model of LiDAR, so that the $\overline{THW_m}(k)$ can be obtained and displayed on the UI interface in real time; Comparing the $\overline{THW_m}(k)$ with THW_d in Car-following Model online, When $\overline{THW_m}(k)$ is greater than THW_d, the driver is prompted to accelerate, conversely, the driver is prompted to decelerate; Furthermore, the unique collision detection box is used to judge the longitudinal and lateral relative positions of the ego vehicle and the front vehicle simultaneously, and the comprehensive evaluation of the horizontal and vertical directions of car-following behavior is completed, which makes up for the deficiency of car-following model. However, there are still some shortcomings in this paper: ignoring the impact of traffic signs, traffic lights and pedestrians on decision-making. In the future, the existing defects will be improved to provide a more perfect and reasonable basis for intelligent driving decision-making system.

6. References

[1] Bian Xiao-hua. 2012. Research on safety characteristics of vehicle operation and corresponding model based on driving behavior. Qingdao Technological University at Qingdao.

[2] Schreier M, Willert V. 2012. Robust free space detection in occupancy grid maps by methods of image analysis and dynamic B-spline contour tracking. *International IEEE Conference on Intelligent Transportation Systems. IEEE*, (Anchorage, AK, USA, September 16-19, 2012),514-521.

[3] Schreier M, Willert V, Adamy J. 2013. From grid maps to Parametric Free Space maps - A highly compact, generic environment representation for ADAS. J. 36,1(January. 2013), 938-944.

[4] Schreier M, Willert V, Adamy J. 2014. Grid mapping in dynamic road environments: Classification of dynamic cell hypothesis via tracking. *IEEE International Conference on Robotics and Automation*. (Hong Kong, China, 31 May-7 June, 2014),3995-4002.

[5] Schreier M, Willert V, Adamy J. 2016. Compact Representation of Dynamic Driving Environments for ADAS by Parametric Free Space and Dynamic Object Maps. J. *IEEE Transactions on Intelligent Transportation Systems*, 17,2, (February.2016),367-384

[6] Liu X, Cai Z. 2010. Advanced obstacles detection and tracking by fusing millimeter wave radar and image sensor data. *International Conference on Control Automation and Systems.* (Gyeonggi-do, South Korea, October 27-30,2010)1115-1120.

[7] Zhang Shuang-xi. 2013. Research on Obstacle Detection Technology Based on Radar and Camera of Driverless Smart Vehicles. Chang'an University at University at Chang'an.

[8] WEI Lang, TIAN Shun, Chris SCHWARZ, LIU Jing-yu. 2017. A Review of Application of Driving Simulation Technology in R&D of Vehicular Intelligence Technology. J. *Highway and Transportation Research and Development.* 34, 12(December.2017), 140-149.

[9] Hong Feng. 2018. Research on Comprehensive Cognition of Traffic Vehicles and Virtual Test for Intelligent Vehicle. Jilin University at Jilin.

[10] Yu Chen-fei. 2013. Study on Forward Collision Warning Method Adaptive to Driver Characteristics. Tsinghua University at Beijing.

[11] Liao Chuan-jin. 2005. Study on Forewarning Information System of Human-centered Vehicle Active Safety. Chongqing University at Chongqing.

[12] Wang Jian-qiang, Chi Rui-juan, Zhang Lei, Li Ke-qiang, Yu Tao. 2009. Study on Forward Collision Warning-avoidance Algorithm Based on Driver Characteristics Adaptation. *Journal of Highway and Transportation Research and Development.* 26, S1(December.2009), 7-12.

[13] Yu Tao, Liu Jin-han, Wang Jian-qiang. 2010. Study on Forward Collision Warning Algorithm for CV. *Journal of Automotive Science and Technology*. 3(May.2010), 15-20.

[14] Liang Jun, Chen Xiao-bo, Cheng Xian-yi, Li Xiao-yan. 2009. Rear-end Alarm Model Based on MAS and Driver Behavior. *Computer Engineering*. 35,10(May.2009),176-178.

[15] Lin Qing-feng, Cheng Bo, Qu Xiao-lei, Zhang Bo, Xu Shao-bing. 2012.Vehicle Rear-end Collision Warning Algorithm Based on Driver's Braking Behavior. *Automotive Engineering.* 34,3(March.2012), 232-235.

[16] Lu Guang-quan, Cheng Bo, Lin Qing-feng, Wang Yun-peng. 2012. Quantitative indicator of homeostatic risk perception in car following. *Safety Science*. 50,9(November. 2012), 1898-1905.

IOP Conf. Series: Materials Science and Engineering **717** (2020) 012015 doi:10.1088/1757-899X/717/1/012015

Multi-objective Excavation Trajectory Optimization for Intelligent Electric Shovel Based on ROS

Zhonglei Fan[1], Tianci Zhang[1], Jianqiang Qiao[2], Guang Li[2], Yonggang Zhao[1], Xin Wang[1], Wei Sun[1] and Xueguan Song[1]

[1] School of Mechanical Engineering, Dalian University of Technology, Dalian, China
[2] Taiyuan Heavy Industry Co., Ltd, Taiyuan, China

sxg@dlut.edu.cn

Abstract. Electric shovel is a large-scale excavation equipment, which often suffer from discontinuous and inefficient excavation due to complicated mine condition. To address these challenges, a multi-objective optimization strategy for electric shovel is proposed to achieve an efficient, stable and intelligent operation. Firstly, the dynamical model of the electric shovel system is established based on Lagrange method. Then, the minimum energy consumption per volume and minimum relief angle difference are regarded as the optimization formulation considering multiple system constraints. Finally, the strategy is applied to the ROS-based intelligent prototype of shovel to verify the effectiveness. The results show that the multi-objective optimization strategy of the excavation trajectory has great guiding significance for the design of the intelligent electric shovel.

1. Introduction

Large-scale electric shovel is a large construction machinery, which is used to excavate coal, oil sands and other materials after blasting in open pit mines and load them onto excavation dump trucks for transportation. In order to meet the urgent needs of efficient excavation in open pit mines, large-scale and intelligent excavation equipment has become an inevitable trend. In industrial production, the excavation trajectory of conventional electric shovel is quite difficult, and the difference in productivity between different drivers is as high as 50% [1, 2]. The existing large-scale electric shovel relies only on manual experience to complete the excavation task, and does not have intelligent sensing and decision-making ability. It is urgent to build an intelligent system to complete the intelligent upgrade of the electric shovel.

The intelligent research of electric shovel mainly consists of intelligent system and excavation trajectory planning. On the one hand, the excavation method using automatic trajectory planning can reduce the burden of the manipulators compared with traditional excavation. On the other hand, the automatic trajectory planning of the electric shovel can greatly improve the production efficiency and stability. Jud D et al. [3] realized the control of output force and running track of end-effector by using hierarchical optimization method, which effectively improved the precision of excavation process. Kim YB et al. [4] calculated the dynamic performance of the excavator using recursive geometric algorithm, and established a highly robust trajectory optimization algorithm based on the dynamic equation. Yoshida T et al. [5] used the discrete element method to establish a geotechnical model and parameterized the excavation trajectory to obtain the optimal excavation trajectory. Awuah-Offei et al. [6] used the energy consumed by the material with unit mass as the objective function and optimized the lifting and pushing speed of the excavator based on the dynamics. Frimpong et al. [7] used Newton-Euler method to establish a dynamic model, and predicted the best excavation performance based on the virtual prototype simulation. J. Seo et al. [8] introduced a excavation task planner that integrates the operational intelligence of construction planner and skilled operator into the robot control mechanism of automatic excavation system, which updates the excavator parameters according to the three-dimensional model of the working environment and cognitive technology to generate the optimal excavation plan. However, most of the existing researches have dealt with excavation trajectory planning into single-objective deterministic optimization problems such as minimum excavation energy consumption. They are failed to comprehensively consider the trajectory planning from multiple perspectives such as efficiency and stability, so it is difficult to meet the application needs of intelligent excavators in the future.

To address the above issues, a multi-objective trajectory planning strategy with minimum energy consumption and minimum relief angle difference is proposed on the intelligent system of electric shovel, and then the optimal excavation trajectory considering multiple system constraints is optimized by the sequential quadratic programming (SQP) method. The planning strategy and intelligent system are applied to the scaled prototype to realize the intelligent test of the shovel, which provides a theoretical basis for the intelligent sensing, decision-making and driverless driving of the electric shovel.

2. System model

The intelligent electric shovel of this paper is to build the intelligent system on the 1:20 scale prototype of WK-55 excavator, as shown in Figure 1. The electric shovel usually consists of an upper component, a lower component, a working component and an intelligent system. The working component includes the stick, the lifting rope, the bucket, the boom and the balance wheel, which is the main operating mechanism for direct contact with the material and completing the excavation task. The upper assembly includes an electromechanical system consisting essentially of a lift motor, a push motor and a platform rotary motor. The most important functional component of the lower assembly is the track travel device. The excavation process of the electric shovel is mainly driven by the lift motor and the push motor to drive the lifting rope and the stick to complete the excavation work.

Figure 1. Intelligent electric shovel model.

2.1. Intelligent shovel system

With the rapid development of artificial intelligence and robotics, robot systems of different operating platforms have enabled developers to do a lot of repetitive work. As an open source system framework with distributed processing, strong code reusability and real-time communication of node messages, ROS is scalable in building shovel intelligent systems [9, 10].

The intelligent electric shovel system described in this paper is based on the ROS robot operating system and is designed with multiple sensors. Ubuntu and ROS systems are installed in the system to realize the robotization of the shovel. The sensing system acquires the shape parameters of the pile by the 3D Lidar scanner and sends it to the Industrial Personal Computer (IPC) for processing [11]. Then the optimal excavation speed parameters (ie motor control parameters) are calculated by the planning algorithm in the decision system and returned to the IPC, and the task command is send to the execution system by the IPC. The execution system is composed of the singlechip, the motor module and the sensor module to complete the task command. The motor module drives each motor to perform the corresponding actions and the sensor module returns the excavation status data to the IPC.

Moreover, the real-time communication and monitoring function of the electric shovel data flow is realized. The entire intelligent system monitors the real-time status information of the electric shovel excavation process as comprehensively as possible, as shown in Figure 2.

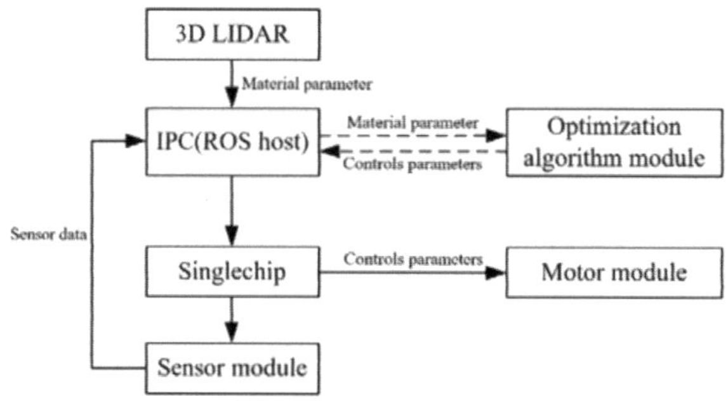

Figure 2. Real-time status of intelligent shovel system.

2.2. Dynamic model

The dynamic model of the working device of the intelligent electric shovel model described in this paper is shown in Figure 3. Suppose τ is the output torque of the working device, M is the inertia matrix of the working device, C is the centripetal force and the Coriolis matrix, and G is the gravity vector. Then the Lagrange equation of the shovel's working device is shown in (1).

$$M(q)\ddot{q} + C(q,\dot{q})\dot{q} + G(q) = \tau \tag{1}$$

The working device of the electric shovel is in a vertical plane only during the excavation process, regardless of the rotary motion of the turret. To simplify the working device into a two degree of freedom manipulator, the dynamic equation can be established as (2) and (3).

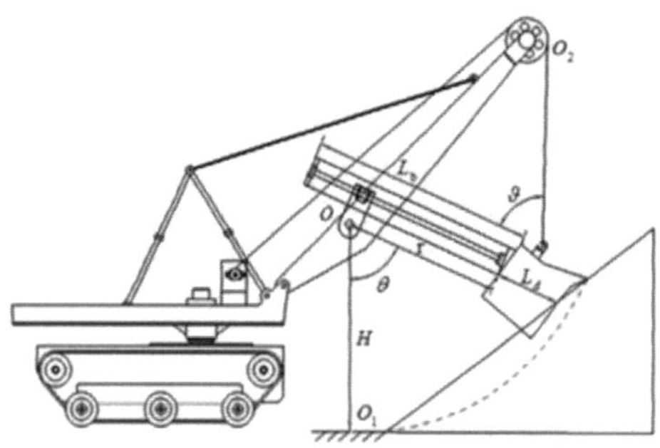

Figure 3. Analysis of working device of shovel

$$\begin{aligned}&[m_b(r^2 - L_b r + \frac{1}{3}L_b^2) + m_d(r^2 + L_d r + \frac{1}{3}L_d^2)]\ddot{\theta} + [2(m_b + m_d)\dot{\theta}\cdot r \\ &-(m_b L_b - m_d L_d)\dot{\theta}]\cdot\dot{r} + [m_b g\sin\theta(r - \frac{L_b}{2}) + m_d g\sin\theta(r + \frac{L_d}{2})] \\ &= F_r\cdot r\cdot\sin\vartheta - F_t\cdot(L_d + r)\end{aligned} \tag{2}$$

$$\begin{aligned}&(m_b + m_d)\ddot{r} - [(m_b + m_d)\dot{\theta}\cdot r - \frac{1}{2}(m_b L_b - m_d L_d)\dot{\theta}]\dot{\theta} \\ &-(m_b + m_d)g\cos\theta = F_h - F_n - F_r\cdot\cos\vartheta\end{aligned} \tag{3}$$

Where m_b and m_d represent the mass of the stick and the bucket respectively, L_b and L_d represent the length of the stick and the bucket respectively, θ represents the rotation angle of the stick relative to the vertical line, and r is the distance that the stick slides during the movement, g represents the gravitational acceleration constant. The main working parameters of the prototype are shown in Table 1 below.

Table 1. Working parameters of shovel proportional prototype.

Parameters/unit	Value	Parameters/unit	Value
V_c/m3	0.068	H/m	0.447
L_b/m	0.520	OO_2/m	0.590
L_d/m	0.165	ρ/kg/m3	1700
w_d/m	0.245	m_b/kg	8
R/m	0.050	m_d/kg	4

3. Multi-objective trajectory optimization strategy

In order to achieve fast, stable and efficient excavation work, based on the Lagrange method to establish the dynamic equation, the minimum energy consumption and the minimum relief angle difference are taken as the objective functions, and the constraints such as the full bucket rate, motion parameters and system physical parameters are considered. Using the point-to-point trajectory planning method, the mathematical model of the excavation trajectory is transformed into an n-order polynomial function by discretizing the time domain [12, 13]. Furthermore, the traditional optimal control problem is changed to directly solve the optimal control parameters, and the optimal excavation trajectory is obtained by solving the optimal control parameters.

3.1. Determination of objective function

a. The minimum energy consumption of a single excavation cycle is taken as the first objective function.

$$f_1(x) = \min \frac{E}{V} \tag{4}$$

$$E = W_r + W_c \tag{5}$$

The first function and the total energy consumption can be formulated as (4) and (5), W_r is the energy consumption of the lifting motor, and W_c is the energy consumption of the pushing motor.

b. The minimum relief angle difference of the excavation process is taken as the second objective function. The second function can be formulated as (6).

$$f_2(x) = \min[\beta_{max} - \beta_{min}] \tag{6}$$

c. The above objective function is weighted and averaged, the multi-objective problem is transformed into a single-objective problem. Then the total optimization objective function can be obtained using (7).

$$f(x) = w_1 f_1(x) + w_2 f_2(x) \tag{7}$$

3.2. Constraints

a. Full bucket rate constraint

The intelligent excavation must at least ensure that the full bucket rate is above 90% and cannot be too large. If the full bucket rate is too large, the motor energy consumption will be increased. Therefore, the maximum limit for selecting the full bucket rate is 110%, as shown in the formula (8).

$$0.9V_c \le V \le 1.1V_c \tag{8}$$

b. Motion parameters constraint

During the excavation operation, the cutting angle of the bucket of intelligent electric shovel changes at a constant time and within a certain range, that is illustrated in (9).

$$\beta_{i\min} \le \beta_i \le \beta_{i\max} \tag{9}$$

The excavation process is mainly carried out by the lifting motor and the pushing motor jointly driving the bucket, and the speed of the pulling rope and the pressing rod should be greater than 0, that is (10).

$$\begin{cases} v_r \geq 0 \\ v_c \geq 0 \end{cases} \tag{10}$$

c. Physical parameters constraint

The effective radius of the pushing process should be limited by the maximum size of the stick, which is the formula (11).

$$r \leq L_b \tag{11}$$

In order to ensure the smooth completion of the excavation process, the bucket teeth should be higher than the height of the pair after the excavation work is completed, which is the formula (12).

$$H \leq y_f \tag{12}$$

Through the dynamic model of the shovel robot established in Section 2.2, the sequential quadratic programming method (SQP) is used to solve the nonlinear problem with multiple constraint conditions, and the trajectory planning of the shovel robot along the given polynomial curve equation is realized. The SQP method is a solution to the constrained optimization problem. It is effective for solving the nonlinear optimization problem, and it has good convergence, high computational efficiency and strong boundary search ability. It is more prominent than other optimization algorithms in real-time trajectory planning, and is suitable for intelligent systems.

4. Simulation and experimental results

Take the scale of the smart shovel as an example. The operating system is Ubuntu16.04 and ROS (kinetic). The intelligent system and the excavation trajectory multi-objective optimization strategy were applied to the proportional prototype to carry out the experiment. The sampling frequency was 10 Hz. At the same time, the vibration of the boom in different directions during the excavation process was described by the data of the three-axis accelerometer in the sensor module. The specific experimental process is shown in Figure 4, which mainly includes the initial position, excavation process, rotation and unloading process, and finally returns to the initial position. The material is made of general-purpose loose sand.

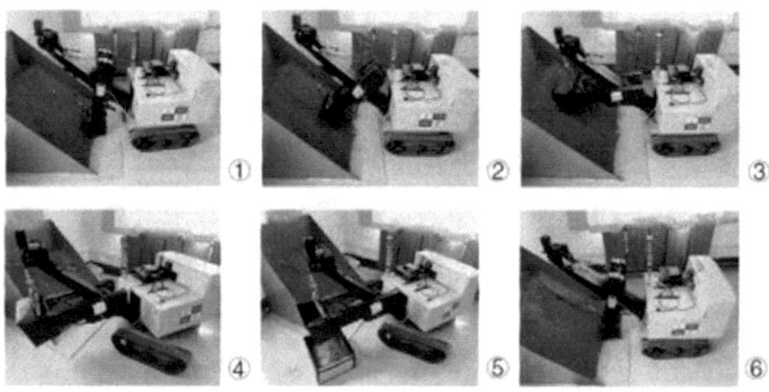

Figure 4. Experimental process of intelligent prototype.

4.1. Comparison of manual and automatic test results

The scale prototype of the intelligent electric shovel studied in this paper is designed into two modes of operation: manual excavation and automatic excavation. The manual excavation process is performed by the operator through the remote control handle to control the lifting motor, push motor and other components. Automatic excavation is to first sense the external environment through laser radar, and then calculate the corresponding excavation parameters by internal decision algorithm to control the corresponding motor to perform the excavation operation.

Under the premise that the operator can skillfully carry out the excavation work, the excavation tests in different modes are performed on the piles of the same pile angle. As can be seen from Figure 5, even the skilled operators have certain differences in each excavation, and the automatic excavation mode using the optimal trajectory planning strategy is basically the same and the excavation process is smooth and stable. Figure 6 shows that for the manual mode in the excavation process, the lifting motor power is not much different from the automatic mode, but the manual mode is much larger than the automatic mode for the power of the push motor. This result may be due to the difficult operation of the linkage of the joystick during manual excavation. In theory, in the working process of the extra-large excavation electric shovel, the power of the lifting motor is greater than that of the pushing motor, indicating that the lifting rope plays a major role in the operation process, the results of the automatic excavation test are consistent with the theoretical results. As can be seen from Figure 6, the automatic excavation process time is about 12 s, and the manual excavation process completed by the skilled operator is about 15 s and it is difficult to maintain consistency, which indicates that the excavation efficiency of the automatic excavation is high.

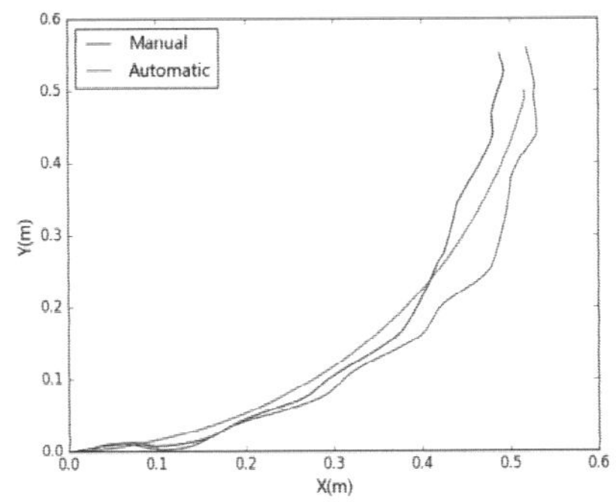

Figure 5. Excavation trajectory of manual and automatic mode

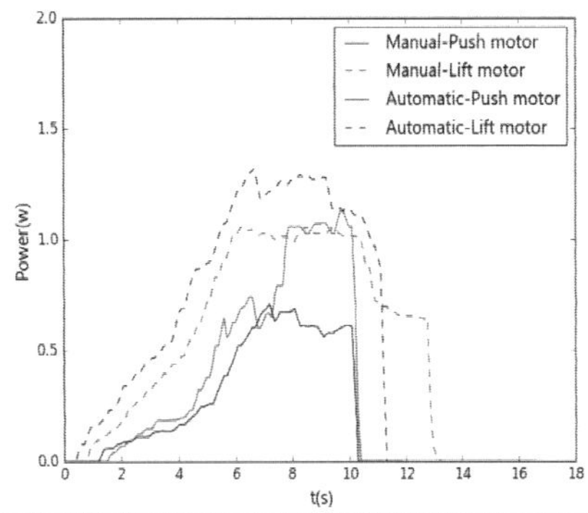

Figure 6. Push and lift motor power changes under manual and automatic mode

Table 2. Analysis of results in manual and automatic mode.

Mode	Full bucket rate	Energy consumption per volume(kJ/m^3)	Std of accelerometer(m/s^2)		
			X	Y	Z
Manual	0.85	29.72	2.59	3.35	0.29
Automatic	0.95	25.10	2.43	2.66	0.37

The experimental results from Table 2 show that the excavation energy consumption of excavation mode with optimization strategy is lower, and the energy consumption per unit volume excavation in automatic excavation process is reduced by 18.41% compared with manual excavation. Manual excavation has a large difference in the full bucket rate due to the operator's human factors, and the average full bucket rate is 0.85. The average value of the automatic excavation full bucket rate is 0.95. Obviously, the automatic bucket mode has a higher full bucket rate. The vibration accelerations of the X, Y and Z axes in the automatic excavation mode are smaller than the manual excavation, which can effectively reduce the vibration of the shovel boom. The reason may be that the control of the motor in the automatic excavation mode is more stable and reduces the excavation. The impact of the impact load during the process.

4.2. Results of single-objective and multi-objective experiments

When w2=0, w1=1, this is the single-objective optimization problem that objective function is determined only by the minimum energy consumption per unit volume of material. When both w1 and w2 are not 0, this is a multi-objective optimization problem that considers the minimum energy consumption and the minimum relief angle difference. In this paper, when considering the influence of excavation time, vibration and other factors, the excavation trajectory is processed into a multi-objective optimization problem, and a sequential quadratic programming method is introduced to optimize the coefficients of polynomial excavation trajectory and obtain the optimal excavation trajectory. Then, the multi-objective and the single-objective optimization problem are respectively applied to the test of the intelligent shovel prototype.

Figure 7 and Figure 8 above show the theoretical and actual excavation trajectory curves for single-objective and multi-objective optimization under different pile-up angles. It can be seen that the actual test results are consistent with the theoretical results, and the excavation trajectory is steeper as the angle of the pile is increased.

Taking the test results of the material with a stack angle of 40 degrees as an example, Figure 9 shows that the single-objective and multi-objective optimization of the motor power trend is consistent, but the instantaneous power of multi-objective optimization strategy is slightly larger than that of single-objective in the middle of the excavation process. There is little difference between the start and end stages of excavation.

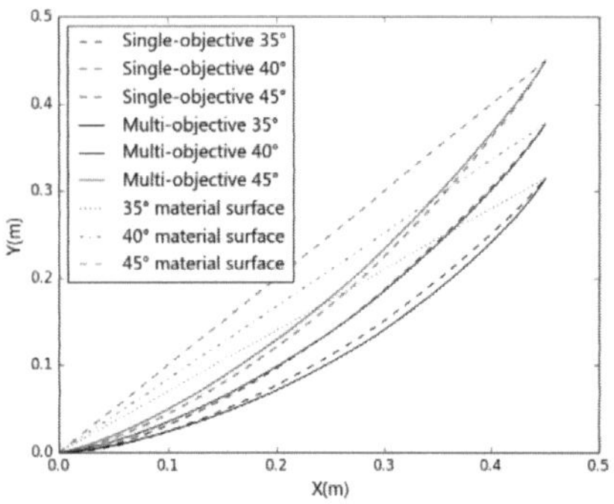

Figure 7. Single-objective and multi-objective test mining trajectory under different angles.

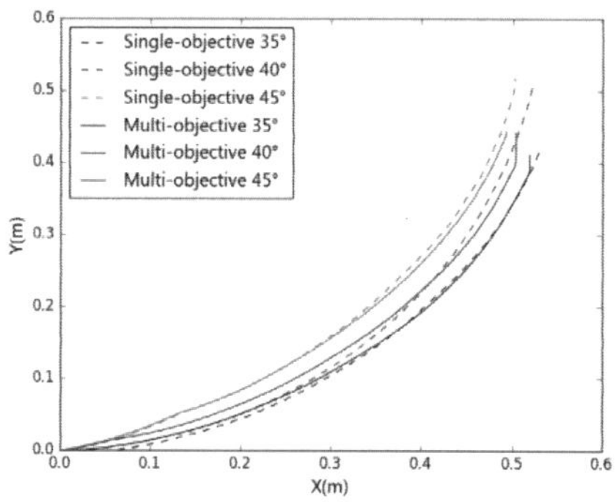

Figure 8. Single-objective and multi-objective test mining trajectory under different angles.

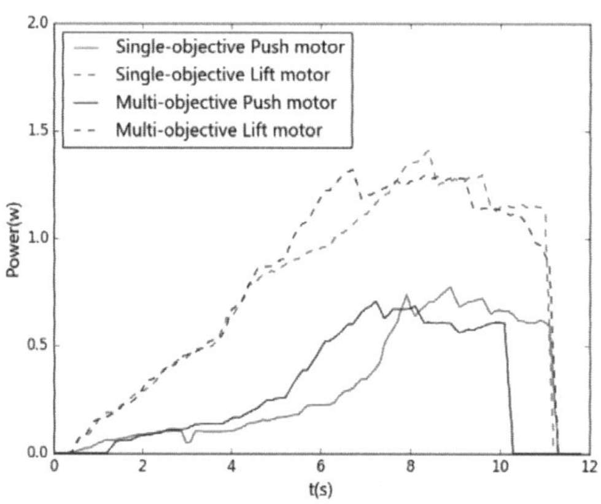

Figure 9. Power change of push and lift motors under the single-objective and multi-objective mining progress

The data in Table 3 shows that the energy consumption per unit volume of the multi-target strategy is slightly higher than the single target. However, the maximum post-excavation angle difference of the multi-target strategy is 0.38 rad, which is smaller than the single-target strategy, and the stability of the excavation process is improved to some extent. The stability can be expressed by the vibration data of the boom. Under the multi-objective strategy, the standard deviation of the acceleration of the boom in the X and Y directions is smaller than that of the single target strategy, and the Z direction is basically the same, indicating that the multi-objective optimization strategy can reduce the vibration of the boom during the working process.

Table 3. Analysis of excavation test results under single objective and multi-objective optimization

Mode	Relief angle difference(rad)	Energy consumption	Std of accelerometer(m/s2)

		per volume(J/m3)	X	Y	Z
Single-objective	0.47	24.14	2.51	2.93	0.34
Multi-objective	0.38	25.10	2.43	2.66	0.37

Due to the practical environment and complicated working conditions of the large electric shovel, the whole machine or local abnormal vibration often occurs, which may easily cause wear failure or fatigue failure of the bearings, gears and other transmission parts, and even broken shaft, boom bending and other serious accidents. It is necessary to reduce vibration of the shovel. Considering the energy consumption and vibration factors comprehensively, although the energy consumed by the single-objective strategy is lower, the vibration is larger. Relatively speaking, the multi-objective strategy is more conducive to prolonging the life of the gear rack, bearing and other equipment of the machine. It is beneficial to improve the performance of the whole machine.

5. Conclusion

Taking the proportional prototype of ROS-based intelligent electric shovel as the research object, a multi-objective optimization strategy based on the minimum energy consumption per volume and the minimum relief angle difference is proposed and tested. This strategy reduces the energy consumption of the whole system. It ensures the stability of the excavation process, and reduces vibration to extend the life of the machine and improves the performance of the whole machine.

(1) Through the optimal control of the intelligent electric shovel excavation trajectory, not only the shovel intelligent decision-making excavation process can be realized, but also the energy consumption of the whole process can be greatly reduced, and the excavation efficiency and operation safety can be significantly improved compared with the traditional manual operation.

(2) Comparing the multi-objective and single-objective optimization strategies of the excavation trajectory, the energy consumption and vibration factors are comprehensively integrated, and the stability of the excavation process under the multi-objective strategy has been greatly improved, which is more conducive to extending the life of the whole machine.

(3) Changing the weighting coefficient and physical constraints of the multi-objective optimization objective function, the optimization objective can be adjusted for different complex working conditions to adapt to the intelligent operation of the electric shovel in the practical environment.

6. References

[1] Patnayak S, Tannant D, Parsons I, Del Valle V, Wong J. Operator and dipper tooth influence on electric shovel performance during oil sands excavation [J]. Int J Min Reclam Environ. 2008.22(2):120–145.

[2] Babaei Khorzoughi, M. and Hall R. A Study of Excavation Productivity of an Electric Rope Shovel for Different Operators [J]. Minerals, 2016. 6(2): p.48.

[3] Jud, D., Hottiger, G., Leemann, P., Hutter, M. Planning and control for autonomous excavation [J]. IEEE Robotics and Automation Letters, 2017,P(99), 1-1.

[4] Kim, Y. B., Ha, J., Kang, H., Pan, Y. K., Park, J., Park, F. C. Dynamically optimal trajectories for earthmoving excavators[J]. Automation in Construction, 2013, 35(2):568-578.

[5] Yoshida, T., Koizumi, T., Tsujiuchi, N., Chen, K., and Nakamoto, Y. Establishment of Optimized Excavation Trajectory for Hydraulic Excavator [J]. Topics in Dynamics of Civil Structures, 2013, Volume 4: 543-554.

[6] Awuah-Offei K, and Frimpong S. Cable shovel excavation optimization for energy efficiency [J].Mechanism and Machine Theory, 2007, 42(8):995-1006.

[7] Frimpong, Samuel & Hu, Yafei. Intelligent Cable Shovel Excavation Modeling and Simulation [J]. International Journal of Geomechanics.10.1061/ (ASCE) 1532-3641(2008)8:1(2).

[8] Seo, J., et al., Task planner design for an automated excavation system [J]. Automation in Construction, 2011. 20(7): 954-966.

[9] Cheng, Y. and D. Shi, ROS Based Remote Robot Task Monitoring and Control System [J]. ITM Web of Conferences, 2017, 12: p. 01023.

[10] Besseghieur, K.L., et al., Trajectory tracking control for a nonholonomic mobile robot under ROS

[J]. Journal of Physics: Conference Series, 2018, 1016: p. 12008.

[11] M.J. Chae, G.W. Lee, J.Y. Kim, J.W. Park, M.Y. Cho, A 3D surface modeling system for intelligent excavation system [J], Automation in Construction. 2011, 20 (7):808–817.

[12] Huang MS, Hsu YL, Fung RF. Minimum-energy point to point trajectory planning for a motor-toggle servomechanism [J]. IEEE/ASME Transactions on Mechatronics, 2012, 17(2):337–344.

[13] Wang, X., et al., Energy-minimum optimization of the intelligent excavating process for large cable shovel through trajectory planning [J]. Structural and Multidisciplinary Optimization, 2018, 58(5):2219-2237.

Acknowledgments

The research is supported by National Natural Science Foundation of China (Grant No.U1608256).

IOP Conf. Series: Materials Science and Engineering **717** (2020) 012016 doi:10.1088/1757-899X/717/1/012016

Prediction of Marine Diesel Engine Combustion Characteristic in Transient Condition By Using Seiliger Process Cycle

Myo Htet Sein and Sun Jian Bo

Marine Engineering College, Dalian Maritime University, China

myohtetseinnord@gmail.com

Abstract. Modern diesel engine cycle can be predicted by applying variety of simulation methods. These models are used for optimization of engine parameters, investigation of control strategies or evaluation phenomena that are difficult to measure. In practice, using engine cycle simulation methods, the number of expensive experiments studies could be reduced and very useful result could be obtained for prediction of engine design parameters in a short period of time. Furthermore, these models are also used to test different solutions (technologies) or control strategies which can help to reduce fuel consumption, increase power output, reduce emissions to meet emission requirement. In diesel engine modeling process, in cylinder pressure measurement is the fundamental method to investigate the combustion process and an effective way to get insight into the combustion phenomena. The main objective of the paper is focusing on the transient modeling of marine two stroke diesel engine and calculating (prediction) the parameter of the cylinder process model by using the Seiliger cycle process. The research approach is carried out using data from a MAN-10L90MC diesel engine and mathematical model of the test engine is calculated by using Microsoft Visual 6.0 VC^{++} programming.

1. Introduction

Today Diesel Engine has been used in different forms and different standards, starting from normal household car to heavy truck, electricity production plant to marine sea going vessels. The most attractive feature of the compression ignition diesel engine is its excellent fuel efficiency, which can surpass 40% in vehicular application and even 50% in large two stroke units of marine propulsion. Due to the above reason, Diesel engine has been used in marine industry as Main engine for many years. The marine industry has always favored using diesel engines for its benefits mentioned.

IOP Conf. Series: Materials Science and Engineering **717** (2020) 012016 doi:10.1088/1757-899X/717/1/012016

However, the regulation is still revolving around the benefits of the players in the industry and is lagging behind other industries. Recently strong regulation has entered into force in attempt to reduce the emission of NOx and SOx by IMO (International Maritime Organization). Especially, emission from transient load conditions are not yet on the table of discussion yet but it is obvious to predict that it is the matter of time in a relatively short term that more strict legislation will come into force to parallelize with other industries. In the other hand, IMO's day by day very strict rules and regulations leads the requirement in improvement of marine diesel engine in very important condition. So it is same as to do experiment in according to comply with MARPOL and to do more in combustion process of Diesel engine to reduce NO_x, SO_x gases and to know better about combustion inside the cylinder. Therefore to overcome this kind of difficulties is by using high technology of modern era high performance computer and it associated programs. There are a lot of advantages by using high performance computers such as saving a lot of money and time, can observe diesel engine's phenomenon in very comfortable way, and get more accuracy and accurate data close to real life Engine. By this way, we can achieve the new modern engines which are complying with IMO's newest rules and regulations.

1.1. Literature review

A number of thermodynamic models of marine diesel engines have been developed for various applications due to the uniqueness of marine diesel engines and their operation and the computer programs for marine applications must be specifically designed, requiring that each application requires a different model. Miedema and Lu (2002) modeled the diesel engine based on the Seiliger (thermodynamic) process. The result of the model shows that the diesel engine behaves like a second order system when operating governor area and more like a first order system in the constant torque (overload) area [1]. Ritzen (2003) introduced a four state mean value model, to make the model usable and simple it was simulated with a fixed step size solver. Chesse (2004) presents a marine diesel engine simulation code designed for real time performance. The various equations are derived from the laws of thermodynamic, thus guaranteeing qualitatively accurate predictions and allowing for easy use with any type of engine. The code is based on the "filling and emptying" method with various simplifications to achieve real time performance. Benvenuto (2005) has applied simulation techniques to marine propulsion systems obtaining interesting results. The transient operation of the main components of marine propulsion plants and their different elements such as engine, shaft lines, clutches, gears, bearings and elastic couplings, have been studied and simulated according to the operational requirements of the ship. Theotokatos (2007) carried out the mapping of the performance and emission parameters of a merchant vessel propulsion system over the ship operating envelope by using a cycle mean value engine model. Yu Ding (2011) investigated the 4 stroke engine combustion process and heat release measurement with Seiliger process and Vibe heat release model [2].

2. Modeling the in-cylinder process with Seiliger model

2.1. Model description and measurement description

The Seiliger process is a simple method to model the in cylinder process of diesel engine. It was use to explore the future of diesel engine in earlier years. By given the trapped pressure and temperature it is capable of catching the main features of the cylinder process, i.e. heat input, indicated work output, maximum pressure and temperature. The Seiliger parameters are applied in a mean value simulation model as the combustion parameters to give a global description of the combustion process. The marine diesel engine in cylinder process can be parameterized by finite stages in different kinds of Seiliger process, providing a simple and reliable method for mean value modeling of the cylinder process. Often the 5 points Seiliger process with two combustion stage is used but in order to model combustion adequately is deemed necessary here [3].

2.2 Six point Seiliger cycle

Figure 1 shows the six points Seiliger process model. The stages can be described as follows;
1-2: polytropic compression, 2-3:isochoric combustion, 3-4:isobaric combustion,4-5:isothermal combustion and expansion, 5-6: polytropic expansion indicating a net heat input caused by late combustion during expansion (advanced seiliger cycle).Stage (6-1) is assumed that the exhaust gas is directly replaced by intake air. In the actual simulation model, the Seiliger process breaks off in point 6, followed by the gas exchange model. [3].

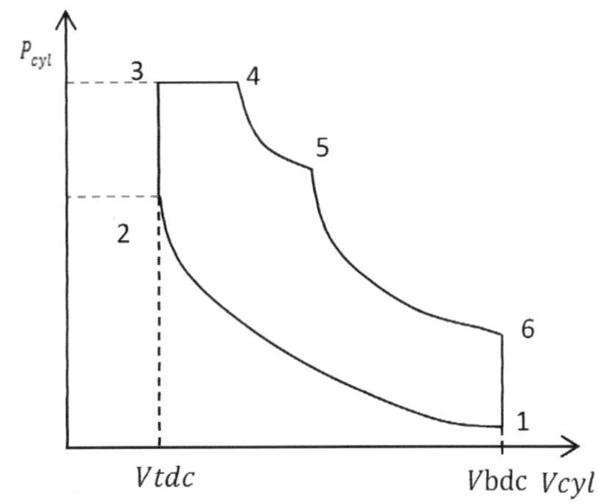

Figure 1. Six Point Seiliger cycle

2.3 Parameterization of the in-cylinder process

The Seiliger process can be described by a finite number of parameters which fully specify the process together with the initial (trapped) condition and the working medium properties. The definition of the Seiliger stages and the Seiliger parameters are given in Table 1.

Table 1. Work done definition and calculation in Seiliger process

Seililger State	Work W[J]
1-2	$W_{12} = \frac{m_{12}R_{12}}{n_{comp} - 1} * (T_2 - T_1)$
2-3	There is no work done between these two points.
3-4	$W_{34} = m_{34}R_{34} * (T_4 - T_3)$
4-5	$W_{45} = m_{45}R_{45}T_4 \log\left(\frac{V_5}{V_4}\right)$

5-6	$W_{56} = \frac{m_{56} R_{56}}{n_{\exp} - 1} * \left(T_5 - T_6\right)$

Once all the Seiliger parameters are obtained, the pressures, temperatures and work done in the various stages of the Seiliger cycle can be calculated according to following Table 1 and Table 2. The polytropic exponents n_{comp} and n_{exp} play an important role in determining the stage 1-2 and 5-6, so they have to be chosen according to experience. [4]

Table 2 Seiliger process definition and parameters

Seiliger Stage	Seiliger Definition	Parameter Definition	Parameter Definition	Seiliger parameter
1-2	$\frac{P_2}{P_1} = r_c^{n_{comp}}$	$\frac{V_1}{V_2} = r_c$	$\frac{T_2}{T_1} = r_c^{n_{comp}-1}$	n_{comp}, r_c
2-3	$\frac{V_3}{V_2} = 1$	$\frac{P_3}{P_2} = a$	$\frac{T_3}{T_2} = a$	a
3-4	$\frac{P_4}{P_3} = 1$	$\frac{V_4}{V_3} = b$	$\frac{T_4}{T_3} = b$	b
4-5	$\frac{P_4}{P_5} = c$	$\frac{V_5}{V_4} = c$	$\frac{T_5}{T_4} = 1$	c
5-6	$\frac{P_5}{P_6} = r_e^{n_{\exp}}$	$\frac{V_6}{V_5} = r_e$	$\frac{T_5}{T_6} = \left(\frac{r_e r_c}{b.c}\right)^{n_{\exp}-1}$	$n_{\exp}, r_e$

3. Experimental investigation of the Seiliger combustion parameters

In this section, the investigation of the Seiliger combustion parameters is carried out which are based on the marine diesel engine measurement. These parameters can be calculated based on the test data of MAN 10L90MC Engine. The main specifications of the MAN 10L90MC Engine are shown in following Table 3.

Table 3 Main specification of marine diesel engine MAN10L90MC

Parameter	Unit	Value
Nominal Engine speed	rpm	84.0
Maximum effective power	KW	43430
Cylinder Number	-	10
Bore	m	0.9
Stroke	m	2.9
Compression ratio	-	16.8
Connecting rod length	m	3.5
Exhaust valve open	deg	114

3.1 Engine Combustion parameters simulation

As described above Marine Diesel engine's combustion is a complicated process and it have to consider several facts and factors from different point of view. To simulate the dynamic characteristics of the diesel engine, several engine parameters must be considered and used. Diesel engines operate with an air excess ratio and it is an important parameter for combustion, engine power, specific fuel consumption, and emissions. It indicates the ratio of amount of scavenge air to amount of actual air for combustion [5-6].

$$\infty = \frac{m_{in}}{m_f \lambda \varnothing_s} \quad (1)$$

$\propto$ = excess air ratio, $\emptyset_s$ = coefficient of scavenging, m_{in} = air mass flow, λ =14.3 for complete combustion of 1 kg of fuel, m_f= fuel mas flow

From above equation, the air fuel ratio Lambda is obtained.

$$\lambda = \frac{m_{in}}{m_f \infty \varnothing_s} \quad (2)$$

Calculation of fuel mass flow,

$$m_f = \frac{mN_{cyl} n_e}{60 N_{st}} \quad (3)$$

N_{cyl}= number of cylinder, n_e= engine rpm, N_{st}=number of stroke, 1 for 2 stroke engine and 2 for 4 stroke engine, m= mass of fuel

The ignition delay of fuel spray is important from the viewpoint of preparing the fuel before injecting into the engine as well as selecting optimum injection.[7].

$$\tau_i = c_1 P_g^{-c_2} \exp\left(\frac{Ea}{R_{mol} * T_g} \right) \varphi^{-c_3} \quad (4)$$

φ= Fuel air equivalence ratio, P_g= integrated mean gas pressure, T_g= integrated mean gas temperature, R_{mol} = the universal gas constant, c_1 = Arrehenius exponent (17.68ms), c_2,c_3 = constant depending on the fuel and injection characteristic (1.7,3.402), Ea = B/CN +25 , B = a constant (618,840-1310,000), CN = cetane number (45……50)

Specific heat at constant volume is noted as c_v and specific heat at constant pressure is noted as c_p as well. It can be defined as follows, In addition to the above-cited quantities; a few other characteristic quantities are used [8-9].

$$specific\ heat\ ratio = \frac{c_p}{c_v} \quad (5)$$

In addition to the above-cited quantities, a few other characteristic quantities are used in modeling of diesel engine which are as follows. The average piston speed is a characteristic speed for internal combustion engines which are consider in calculation of heat coefficient.

$$piston\ speed = \frac{2.L.n_e}{6o} \quad (6)$$

The compression ratio is the total cylinder volume in reference to the compression volume.

$$r = 1 + \frac{V_d}{V_c} \quad (7)$$

3.2 Seiliger process calculation and combustion parameter calculation

In any given case, the trapping volume, V, is a constant. This is also true for the gas constant, R, for gas at the prevailing gas composition at the trapping point. The gas constant for exhaust gas, Rex, is almost identical to the value for air, Ra. Because the cylinder gas composition is usually mostly air, the treatment of R as being equal to Ra invokes little errors. For any one trapping process, over a wide variety of scavenging behavior, the value of trapping temperature, T_{tr}, would rarely change by 5%. Therefore, it is the value of trapping pressure, P_{tr}, that is significant variable in consideration of trapping mass. The net effect of the cylinder scavenge process is to fill the cylinder with a mass of air, ma, with in a total mass of charge, m_{tr}, at the trapping point. [10].

Basically, two thermodynamic laws are used in cycle analysis, the ideal gas law and the polytropic compression law. The mass of gas in a combustion chamber remains essentially constant during the compression process. If the temperature also remained constant, Equation (9) indicates that the pressure would vary inversely with the volume. In an engine, the temperature increases considerably during the compression stroke; therefore, the pressure increases more than in an isothermal process. The initial pressure and temperature of the process represent (P_1 and T_1) a scavenge air pressure and temperature of the engine inlet which can be referred from engine manual. The volume at trapped condition is V_1 can be calculated by using the crank angle (α), assuming a constant engine speed to calculate the crank angle in following equation（16）[11];

$$V(\infty) = A_b.L_s\left[\frac{1}{r-1}+\frac{1}{2}\left(\left(1+\cos\infty\right)+\frac{1}{\lambda}\left(\sqrt{1-\lambda^2}.\sin^2\infty\right)\right)\right] \tag{8}$$

After obtaining the all Seiliger parameters and pressures, temperatures of the respective states, the work done produced for any given engine cycle and the effective pressure of the simulated diesel engine can be calculated as follows.

$$W_i = \int_0^{360} P_{cyl}(\infty)dV(\infty) \tag{9}$$

$$P_e = \eta_m \cdot P_i(\infty) \tag{10}$$

Finally Power of the marine diesel engine is simply calculated for N rpm and n cylinder number.

$$EnginePower = P_e LANn \tag{11}$$

4. Result of the simulation

The result compared with the bench test data shows that the model can reflect the actual system's steady-state and dynamic characteristics correctly. Following figures show the comparison of actual result of the test engine and simulation result of the Seiliger process. Also the important characteristics in combustion of marine diesel engine such as maximum pressure, power, and specific fuel consumption and compression pressure are shown in Table (4) and (5) with errors compare with the actual ship trial test results. From which we can said that the errors are less than 5% and which are within the acceptable region. It is clear that the diesel engine combustion process is extremely complex and some important consequences can be note down. A higher engine compression ratio can be used in the diesel engine which improving its fuel conversion efficiency and improving the efficiency of diesel engines at higher loads. The delay period between the start of injection and start of combustion must be kept short.By varying the amount of fuel injected per cycle with the engine air flow essentially unchanged which cause lower pumping work and improving the engine part-load mechanical efficiency. Advancing injection timing or increasing injection pressure improves combustion efficiency raises combustion temperature which leads to higher NOx formation. Also a rise in inlet temperature results in a rise of output temperature as we expected. But inlet temperature has a large effect on lambda which values increases when decreasing inlet temperature. As conclude, the paper simulated working process of marine diesel engine by combining diesel engine and Seiliger process model which can reflect not only some average parameters but also important parameters in real time with high performance and less error.

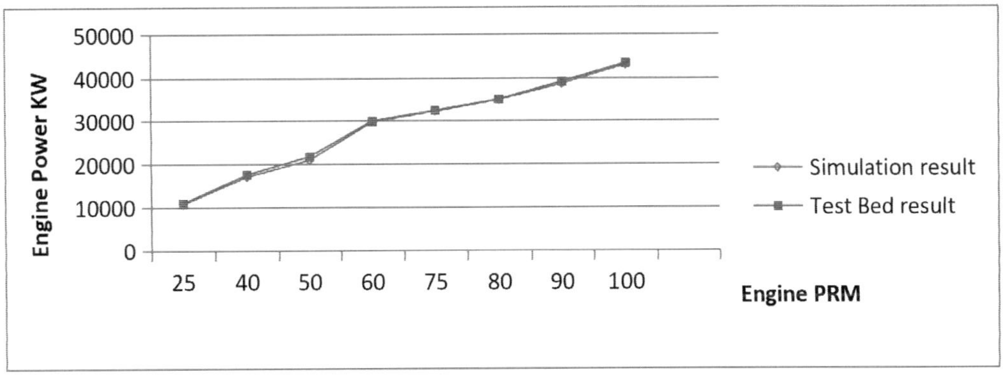

Figure 2. Power Comparison

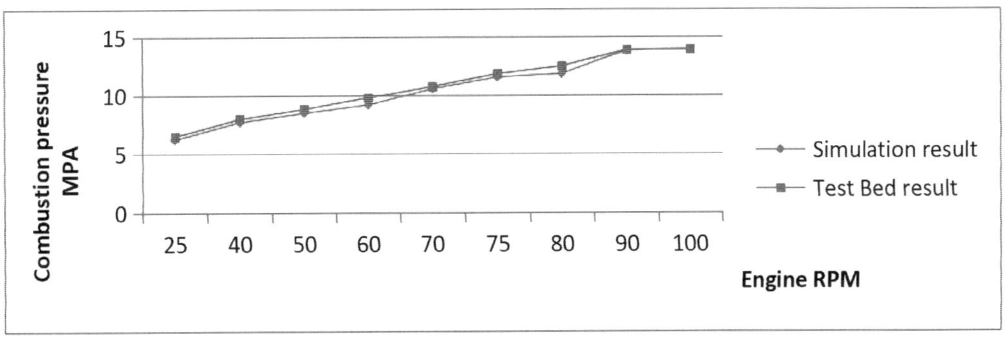

Figure 3. Pmax or Combustion pressure Comparison

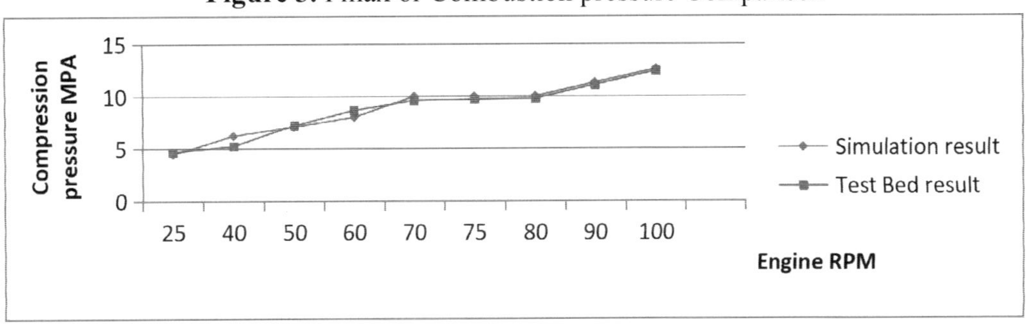

Figure 4. Pcomp Compression pressure Comparison

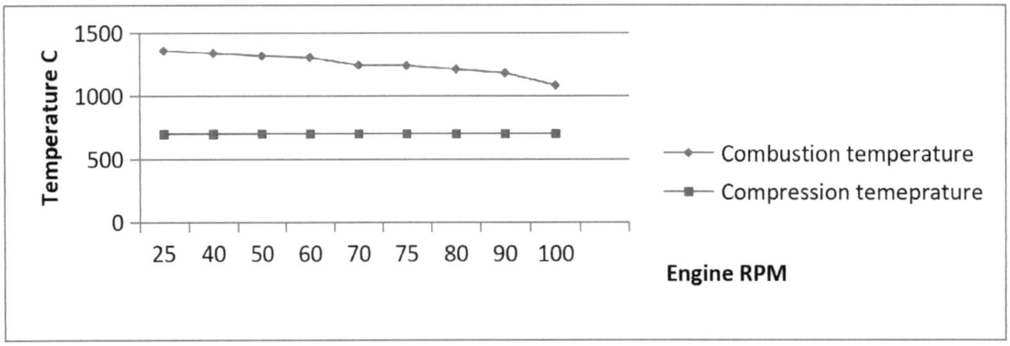

Figure 5. Simulation Result of Combustion Temperature and Compression Temperature

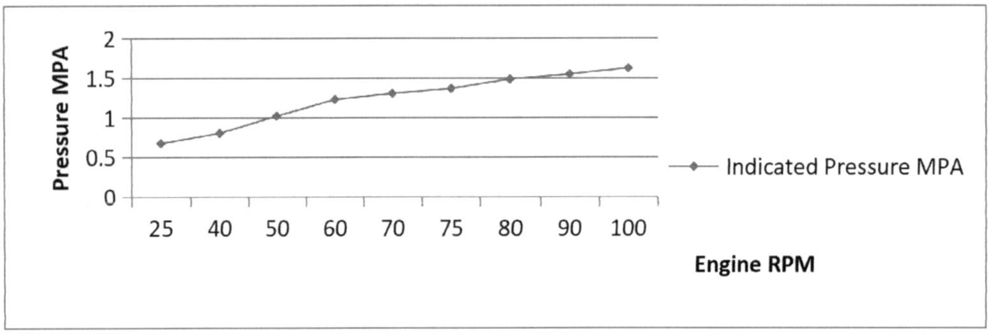

Figure 6. Simulation Result of Mean Indicated Pressure

Table 4. Simulation and Test Bed Result Error

Load %	RPM	Power (KW)			Pmax (MPA)		
		Test	Simulation	Error%	Test	Simulation	Error%
25	51.8	10976	10763.62	1.934912	8.81	6.169	4.9460708
50	65.2	21786	20935.95	3.901822	10.76	8.483	3.7116912
70.4	72.8	30413	29896.8	1.697300	11.84	11.139	3.5223048
75.4	74.6	32493	32304.33	0.580532	13.92	11.56	2.3648648
90.8	79	39122	38485.67	1.626544	13.86	13.88	0.2514367
100.8	84.6	43432	43043.77	0.89387	13.918	13.92	0.4329004

Table 5. Simulation and Test Bed Result Error

Load %	RPM	Pcomp			SFOC		
		Test	Simulation	Error%	Test	Simulation	Error%
25	51.8	7.2	4.425	4.838709	189.5	193.23	1.968337
50	65.2	9.47	7.08	1.666666	180	187.313	4.062777

70.4	72.8	9.76	9.571	1.066525	175.5	181.68	3.521367
75.4	74.6	11.05	9.96	2.049180	175.8	176.82	0.580204
90.8	79	12.36	11.26	1.900452	176	178.9	1.647727
100.8	84.6	13.48	12.56	1.618122	177	179.54	1.435028

5. Conclusion

The simulation is program is now link with other simulator program and provide as teaching assistant tool. The more research have to be done in simulating modern diesel engine with late injection timing, high pressure injection equipment, and even common rail system, with the simulation models which are based on the Seiliger model as basic principle. Finally the combustion model is generally at the center of interest in the marine diesel engine model literature and it is the most complex phenomena occurring during the engine cycle. As conclude, the developed the simulated mathematical model of the marine diesel engine is good compromise for students who are engaged in the new generation of the marine diesel engine development.

6. References

[1] Miedema SA and Lu Z. June 12-15, 2002, *The Dynamic Behavior of a Diesel Engine, Denver*, Colorado, USA Proc. WEDA XXII Technical Conference and 34th Texas A&M Dredging Seminar.

[2] Ding, Y.; Stapersma, D.; Grimmelius, H.T., 2009, *Cylinder process simulation with heat release analysis in diesel engine*, Wuhan, China , In Proceedings of the APPEEC2009.

[3] Ding, Y.; Stapersma, D.; Grimmelius, H.T., 2010,*Characterising Heat Release in a Diesel Engine: A comparision between Seiliger Process and Vibe Model*, Bergen, In Proceedings of the CIMAC Congress.

[4] Ding, Y., 2011, , Delft, The Netherlands, *Characterizing combustion in diesel engines using parameterized finite stage cylinder process models*. Ph. Dissertation, Delft University of Technology, VSD,ISBN; 978-90-6562-289-1.

[5] Gerasimos, P.; Theotokatos, G., 2008, *Ship Propulsion Plant Transient Response Investigation using a Mean Value Engine Model, International Journal of Energy*, vol.2, p 66-74,.Heywood JB. 1988. Internal Combustion Engine Fundamentals. McGraw Hill.

[6] Theotokatos, G.; Tzelepis, V, 2015, *A comuptional study on the performance and emissions parameters mapping of a ship propulsion system*, SAGE Journal, Issue 1, vol 229.

[7] Heywood JB, 1988, *Internal Combustion Engine Fundamentals*, McGraw Hill

[8] G.Stiesch; Merker G.P; Schwarz C; Otto F., 2005, *Simulating Combustion: Simulation of combustion and pollutant formation for engine-development*, Springer.

[9] Guzzella, L.; Onder, C.H, 2007, Berlin, *Introduction to modeling and control of IC engine systems*, Springer.

[10] Ding, Y.; Stapersma, D.; Grimmelius, H.T, 2011, *A new method to smooth the in-cylinder*

pressure signal for combustion analysis in diesel engines, Proc. Ins. Mech. Eng, Part A, vol 225, p 309-318.

[11] Zacharewicz, M.; Kniaziewciz, T., 2017, *Modelling of the operating process in a marine diesel engine*, Journal of Marine Engineering & Technology, vol 16, Issue 4, p 193-199.

Acknowledgments
I would like to acknowledge with gratitude, support and kind of my supervisor Prof. Sun Jian Bo and a number of my friends and colleagues in encouraging me to start the work, preserve with it and finally to publish it. Finally I would like to show my grateful feeling to China Scholarship Council (CSC) who awarded the scholarship to a huge number of foreign students and Dalian Maritime University who give me enormous knowledge in my marine engineering professional.

IOP Conf. Series: Materials Science and Engineering **717** (2020) 012017 doi:10.1088/1757-899X/717/1/012017

Design and Simulation of the Controller for Pneumatic Self-Balancing Hoist System

Liang Yang[1, 2], Peng Yin[1], Rong Yin[2]

[1] School of Mechanical and Automotive Engineering, South China University of Technology, Guangzhou 510640, China
[2] R&D Department, Guangzhou Hyetone Electromechanical Equipment Co., Ltd., Guangzhou 510407, China

Corresponding author's e-mail address: 18814113085@163.com (Peng Yin), leonyounggod@msn.com (Liang Yang).

Abstract. The design of the controller of the pneumatic hoist system is the most important step in the whole work process. The quality of the design of the controller is directly related to the realization of the system. This paper elaborates the detailed design of the system in a few steps. Firstly, a set of optimal control parameters k_p^*, k_i^*, and k_d^* are selected. Then the fuzzy self-tuning PID controller is designed for the valve-controlled cylinder system. Finally, the response characteristic simulation of the fuzzy self-tuning PID controller is designed based on Matlab/Simulink platform. From the result and analysis, it can be seen from the simulation results that through the adjustment of the fuzzy self-tuning PID controller, the system adjustment time t_s are within 0.8s, and the steady-state error rate is controlled within 3%.

1. Introduction

The pneumatic self-balancing hoist system allows the operator to easily carry out various handling, assembly and special positioning tasks and is widely used in industrial fields such as home appliance manufacturing, automobile assembly, and weapon production [1]. It plays an important role in improving production efficiency and improving the working environment. At present, the pneumatic self-balancing hoist system independently developed in most of the country has the disadvantages of low positioning accuracy, poor operation stability, small suspension stroke and complicated pneumatic routes. Therefore, it is necessary to develop a pneumatic hoist system that can be widely used in the production of heavy objects lifting, material handling and full suspension.

The controller design of the pneumatic self-balancing hoist system is one of the important steps of the whole system [2]. The quality of the controller design is directly related to the realization of the system function. The system mathematical model is based on the basic theorems of fluid mechanics, pneumatic transmission, dynamics, etc., and applies mathematical forms to quantitatively describe the static and dynamic relationships between various parameters in the physical model system. In the control system, accurate mathematical models play a very important role in analysing component and system characteristics and design control strategies [3].

2. Design strategy

Using the appropriate control strategy is a key part of the entire control system. The quality of the control strategy is directly related to the operation of the entire system. The pneumatic system is complex because of its non-linear characteristics. Hence the pneumatic hoist control system is much more complicated than other types of servo control systems such as hydraulics.

2.1. PID Controller Design

PID control is one of the earliest developed control strategies [4]. Because of its simple algorithm, good robustness and high reliability, it is widely used in process control and motion control. When the structure and parameters of the controlled object cannot be completely mastered, When the precise mathematical model is not available, it is most convenient to apply PID control technology [5].

The PID controller is a linear controller. The principle is as follows: The deviation is calculated according to the given value *r(t)* and the output *c(t)*, which is expressed as follows [6]:

$$e(t) = r(t) - c(t) \tag{1}$$

By linearly combining the differential, integral, and proportional deviations to form a control quantity, and thus effectively controlling the object, the control law can be expressed by the following equation:

$$u(t) = k_p \left[e(t) + \frac{1}{k_i} \int_0^t e(t)dt + \frac{k_d de(t)}{dt} \right] \tag{2}$$

In the above formula, k_d represents the differential time constant, k_i represents the integral time constant, and k_p represents the proportional coefficient. These three parameters determine the controller performance. Therefore, the core of the design of the PID controller is to determine 3 parameters for controls. According to the mathematical model of the valve-controlled cylinder deduced in the previous section, a system simulation model based on PID control is established, as shown in Figure 1.

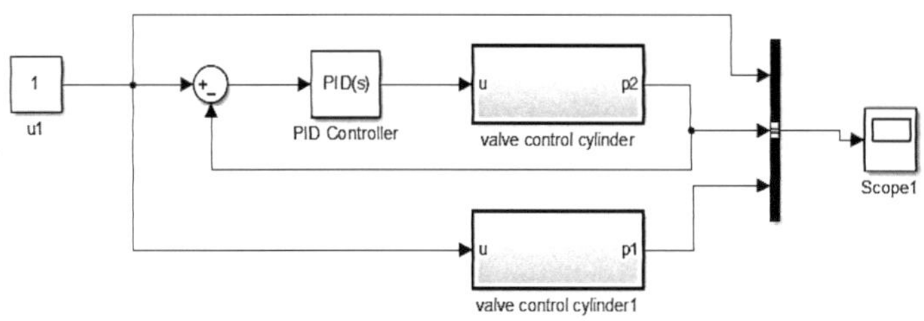

Figure 1. PID control simulation block diagram.

The response characteristics of the output signal to the input signal are simulated under the same mass load, and the optimal control parameters are selected according to the simulation results. Wherein, the input voltage of the electric proportional valve is an input signal, and the pneumatic pressure value in the cylinder is an output signal. In order to ensure the validity of the control parameter selection, three sets of simulations are performed on the PID control system. The input signal settings under different loads are shown in Table 1.

Table 1. Input voltage values under different loads.

No.	*M*/kg	*u*/V
1	25	1.09
2	75	3.27
3	125	5.46

Through a large number of simulation analysis, a set of PID parameters is obtained: when k_p is 1.43, k_i is 0, and k_d is 0.04, the response of the system to the input signal is ideal, the introduction of the PID controller significantly improves the dynamic response performance of the system, reduces the number of system oscillations, shortens the rise time, and reduces the steady-state error. The simulation results also show that the response time of the system is close to 3s after the introduction of

PID controller. It is difficult to improve the response speed by adjusting the three control parameters k_p, k_i, and k_d. Therefore, it is necessary to design a more suitable control strategy to improve the performance of the system.

2.2. Fuzzy Control Principle

Fuzzy Control is an intelligent control method based on fuzzy set theory, fuzzy linguistic variables and fuzzy logic reasoning [7]. It mimics human fuzzy reasoning and decision-making process [8]. The method firstly classifies the operator or the experience into fuzzy rules, then fuzzes the signal from the sensor, and uses the fuzzy signal as the input of the fuzzy rule to complete the fuzzy reasoning[9]. The resulting output after inference is added to the actuator. The basic principle block diagram of fuzzy control is shown in Figure 2.

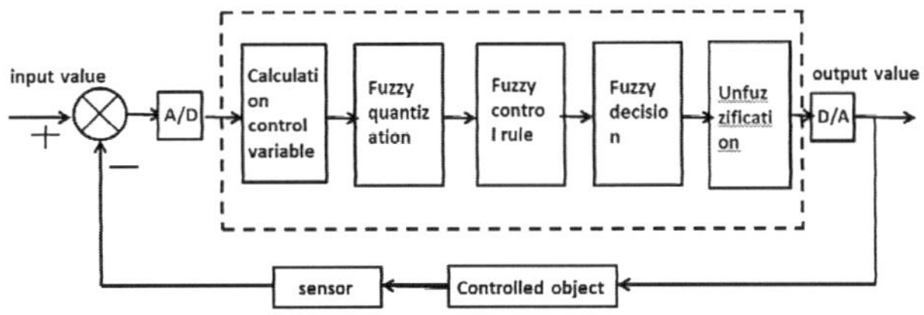

Figure 2. Block diagram of fuzzy control.

3. Fuzzy-PID controller simulation

The Simulink software package in Matlab software can be used to realize the establishment and dynamic simulation of PID control system, fuzzy control system and Fuzzy-PID control system [10]. When designing the fuzzy controller, it includes the determination of fuzzy input and output, the determination of membership function, the setting of the range of input and output, the establishment of fuzzy rule base and the method of defuzzification [11].

3.1. Building a system simulation block diagram

According to Figure 3, composing the PID controller, fuzzy controller and Fuzzy-PID controller in Matlab/Simulink, the results are shown in Figure 3 to Figure 6. Figure 3 is a block diagram of the PID controller model. In the block diagram, kp_0, ki_0, and kd_0 are preset values of the Fuzzy-PID controller. Figure 4 shows the fuzzy controller model. Figure 3 and Figure 4 are packaged and connect to get the simulation block diagram of fuzzy PID control, as shown in Figure 5. The model of Figure 5 is packaged and combined with Figure 3 to obtain the simulation model of the fuzzy self-tuning PID control system of the pneumatic hoist system, as shown in Figure 6.

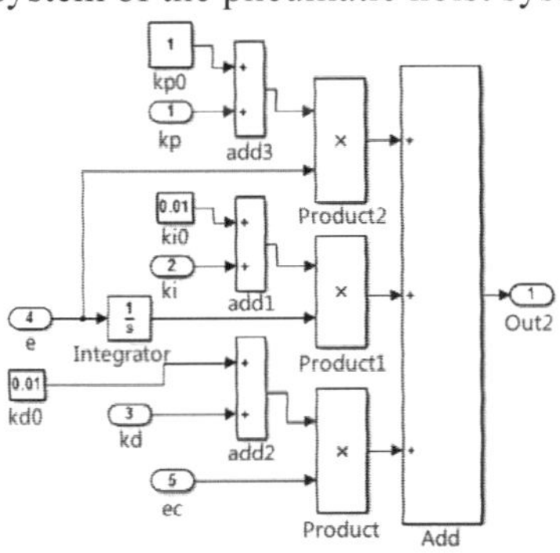

Figure 3. PID controller simulation block diagram.

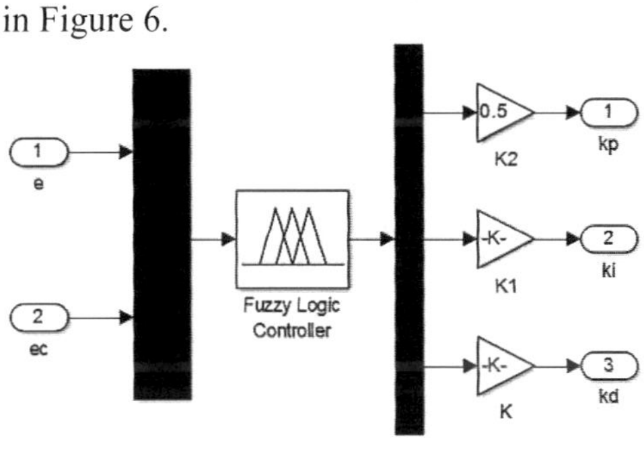

Figure 4. Fuzzy controller simulation block diagram.

3.2. Simulation and result analysis

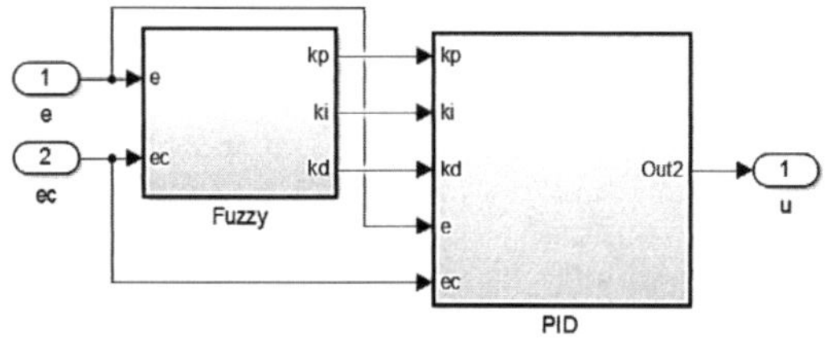

Figure 5. Fuzzy PID controller simulation block diagram.

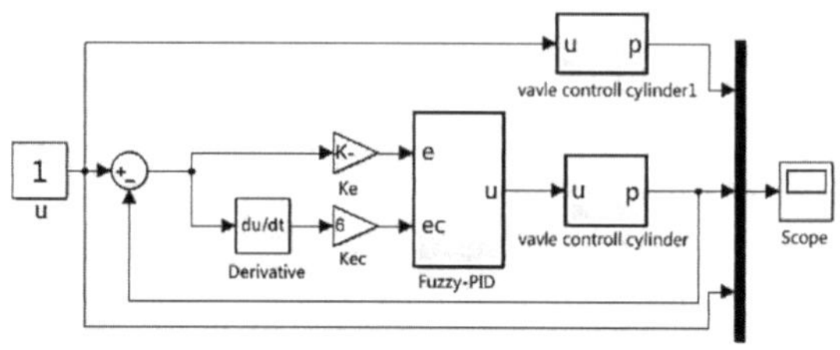

Figure 6. Simulation model of fuzzy self-tuning PID control system for pneumatic hoist system.

The system parameters and simulation parameters are taken in Table 2, and the optimal parameters obtained by the PID controller simulation process are used as pre-tuning parameters of the fuzzy adaptive PID controller. The simulation results are shown in Figure 7, Figure 8, and Figure 9.

Referring to Figures 4, 5, and 6, it can be seen from Figures 7, 8, and 9 that the response speed of the Fuzzy-PID control system is significantly higher than that of the PID control system. Relatively small, and the simulation curve of the Fuzzy-PID control system agrees well with the set curve, and the adjustment time t_s is in the range of 0.5~0.8 s, which also explains the correctness of the controller design. The figure also shows the under each load. The response curve has an error rate of 0~3%, which is difficult to eliminate by adjusting the controller parameters. The possible reasons for the analysis are as follows: In the dynamics module of the valve-controlled cylinder model, the damping coefficient C_s data is derived from the third chapter cylinder friction characteristics. The fitting results of the experimental study, in this experiment, neglected the subtle influence of the cylinder pressure on the damping coefficient, resulting in a small C_s setting. During the simulation process, after the system reaches a steady state, the gas volume *dv/dt* change A in the cylinder is too large. According to the equation (2), the steady state value of the cylinder pressure is small.

For further observation, the bottom of the simulation curve is enlarged, as shown in Figure 10. It can be seen from the figure that the system response curve has a delay at the initial moment, and theoretically, the gas source and cylinder pressure difference is the largest at the initial moment, and the gas mass flow rate, the maximum pressure rises the fastest. The reason for the analysis is as follows: the initial pressure of the cylinder is set to the atmospheric pressure Pa (relative pressure value is 0), and the gas passes through the air pipe from the gas source to the gas pipe. During this time, the air pressure of the cylinder is almost It does not change, so the response curve has a delay and the delay time is close to 0.15 s.

Table 2. Pre-setting parameters table.

Pre-setting parameters	Values
k_p*	1.43
k_i*	0.06
k_d*	0.04

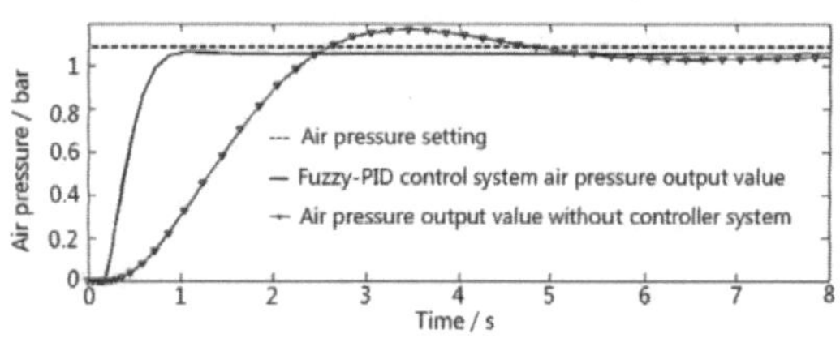

Figure 7. Response curve under 25 kg load.

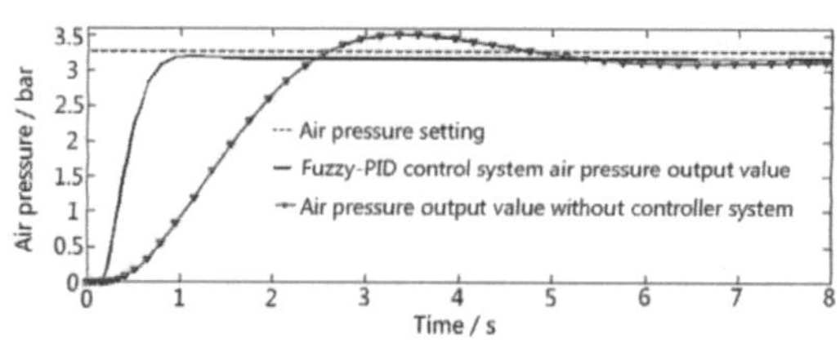

Figure 8. Response curve under 75 kg load.

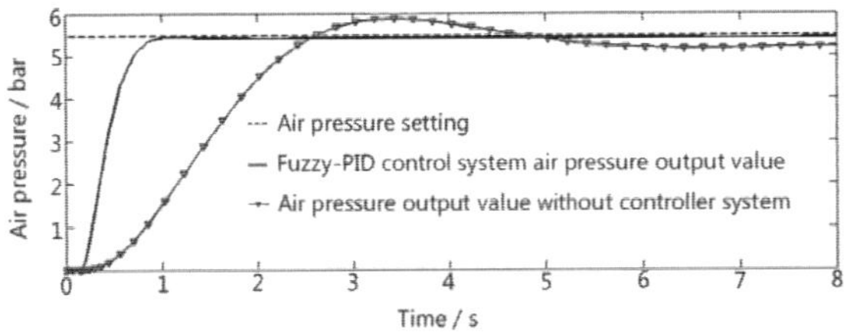

Figure 9. Response curve under 125 kg load.

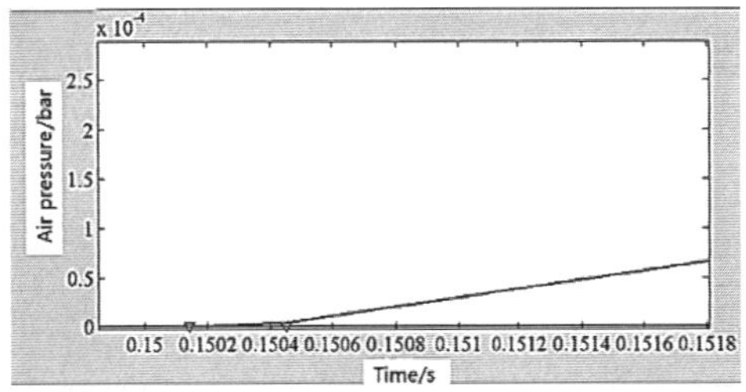

Figure 10. Magnified view of the bottom of the 125 kg response curve.

4. Conclusion

Based on the Matlab/Simulink simulation, the mathematical model of the valve-controlled cylinder system is firstly established based on the flow equation, pressure equation and dynamic equation of the cylinder. Secondly, the PID controller is designed for the valve-controlled cylinder system. Through the simulation on the Matlab/Simulink platform, the dynamic response performance of the system is greatly improved by the introduction of the PID controller. The fuzzy self-tuning PID controller is designed again for the valve-controlled cylinder system. The optimal control parameters k_p^*, k_i^*, and k_d^* obtained by PID control are used as pre-setting values. At the same time, the controller design is detailed and calculated. Finally, based on the Matlab/Simulink software, the response characteristics, simulation and result analysis of the fuzzy controller of the designed fuzzy self-tuning PID controller can be seen from the simulation results. Through the adjustment of the fuzzy self-tuning PID controller, the system adjustment time t_s is within 0.8 s, stable. The state error rate is controlled within 3%.

5. References

[1] Zaguroli Jr J 2008 Electric motor driven traversing balancer hoist. Google Patents
[2] Liu, X., Zhao, R. & Zhao, Z. J, 2012, Automatic lifting system design and application of large load powered support test rig. *Journal of Coal Science and Engineering (China),* Volume 18, Issue 4, pp 428–431
[3] Ching-Han Chen, Chien-Chun Wang, Yi Tun Wang, and Po Tung Wang, 2017. Fuzzy Logic Controller Design for Intelligent Robots. *Mathematical Problems in Engineering*. pp 12
[4] Yao X, Su X and Tian L *2009 International Workshop on Intelligent Systems and Applications,2009),* vol. Series: IEEE pp 1-5
[5] Hassan A A and Domzalski D B 2002 Jet actuators for aerodynamic surfaces. Google Patents
[6] Chao, C.-T.; Sutarna, N.; Chiou, J.-S.; Wang, C.-J. An Optimal Fuzzy PID Controller Design Based on Conventional PID Control and Nonlinear Factors. *Appl. Sci.* Vol. 9, Iss: 6. pp 1224
[7] Ching-Han Chen, Chien-Chun Wang, Yi Tun Wang, and Po Tung Wang, 2017. "Fuzzy Logic Controller Design for Intelligent Robots," *Mathematical Problems in Engineering*, vol. 2017
[8] Alexander M. Veprik, Sergey V. Riabzev, and Nachman Pundak, 2004. Dynamically counterbalanced pneumatically driven expander of a split Stirling cryogenic cooler, *Infrared Technology and Applications XXX*
[9] Y. J. Zhang et al., 2011. Research on Four Type-2 Fuzzy Reasoning Models, *Advanced Materials Research*, Vols. 204-210, pp. 406-411
[10] Soyguder S, Karakose M and Alli H 2009 Design and simulation of self-tuning PID-type fuzzy adaptive control for an expert HVAC system *Expert systems with applications* **36** 4566-73
[11] Cheng X C, Jiang X W and Cheng Z X *Applied Mechanics and Materials,2014),* vol. Series 457*)*: Trans Tech Publ) pp 1503-6

Acknowledgements
This paper is supported by Guangzhou Scientific and Technological Projects (NO.201802010067).

IOP Conf. Series: Materials Science and Engineering **717** (2020) 012018 doi:10.1088/1757-899X/717/1/012018

Research on Voice Print Recognition Algorithm based on Wavelet-GA-BP Neural Network

Kong Yanfeng[1,2] Jia Lianxing[1] and Zhang Jiang[2]

[1] College of Information and Communication, National University of Defense Technology, Wuhan, Hubei, China
[2] 92815 Army Branch of Weapon, Xiangshan, Zhejiang, China

13407160738@163.com; simulab@126.com; 103984675@QQ.com

ABSTRACT. Aiming at the fact that common pattern recognition algorithms such as LPCC and MFCC have flaws in voice-print recognition, the paper put forward a new algorithm, which employed wavelet analysis, BP neural network and niche genetic algorithm. Firstly, the algorithm extracted time-domain and frequency-domain characteristic variation of voice signal by wavelet transform, secondly, trained the neural network niche genetic algorithm, which solved the problem of local minimum value caused by normal multi-layers neural network, lastly, took the wavelet transform variations as the training data of the optimization neural network. Simulation experiment was carried out. The results indicate the algorithm in this paper is superior to current recognition algorithms, it has fast recognition velocity, high recognition rate, low fault rate and strong robustness with different voice-makers, besides, it can automatically correct errors.

1. Introduction

As network technology rapidly develops, the information security is more and more important. Traditional password authentication based on personal identification technology has been insufficient, while biometrics technology is increasingly mature and showing its superiority. Among them, voice print recognition is a new technology developed in recent years, it is simple, precise, economical and non-contacted in comparison with other recognition technology [1].

This paper puts forward a new recognition algorithm based on wavelet analysis and BP-GA optimization algorithm, compared with traditional algorithm such as LPCC and MFCC, it has the advantage of fast recognition velocity, high recognition rate, low fault rate, automatic error correcting, and strong robustness with different voice-makers.

2. Wavelet Transform of Voice Signal

For the object of voice print algorithm is digital variation, the pre-processing is needed for voice signal. The LPCC and MFCC algorithm employ FFT frequency-domain analysis method, which can only extract frequency-domain feature, but can not extract time-domain feature, even if add time-window, the time-domain resolution is still inadequate, and the frequency-domain features are easily lost. Wavelet transform is a new time-frequency analysis method, which has the merit of resolution adjustable, strong anti-interference and can reflect signal's non-stationary and transient feature[2-3]. Besides, it has low computation and is easy to realize, fully reflects not only human ears's feature but also voice signal's dynamic feature, and improve the rate of final voice print recognition.

Suppose one division voice signal $f(t) \in L^2(R)$, its wavelet transform is defined as the inner product of signal and wavelet base function, that is:

$$W_F(a,b) = (f, \psi_{a,b}) = |a|^{\frac{1}{2}} \int_R f(t)\psi\left(\frac{t-b}{a}\right)dt \tag{1}$$

Similar with short-time Fourier transform, the original signal can be re-constructed from its wavelet transform, the formula is:

$$f(t) = C_\psi^{-1} \iint R^2 W_F(a,b)\psi_{a,b}(t)\frac{dadb}{a^2} \tag{2}$$

Suppose the signal with noises is:

$$s(i) = f(i) + e(i), i = 0,1,2,\cdots,n-1 \tag{3}$$

In eq.3, $f(i)$ is the real voice signal, $e(i)$ is gaussian white noise or 0ther noise, $s(i)$ contains noise signal and useful low frequency signal to be extracted. Voice signals are mainly stable signals in engineering, while noises are mainly high-frequency signals. So we should select a wavelet base and determine the layer of wavelet decomposition. Make 3-layer decomposition of signal, the figure is shown in Fig.1

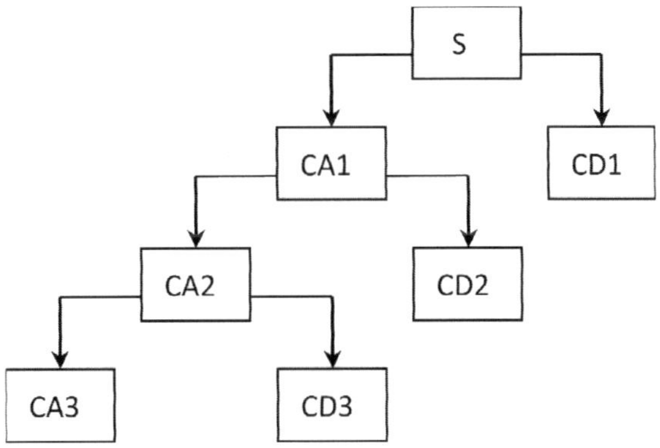

Figure 1. Schematic diagram of wavelet signal decomposition

Mark CD1, CD2, CD3, CA3 as d1, d2, d3, a3 and make normalization, mark the normalized parameters as Q1, Q2, Q3, Q4. Construct a new vector [Q1,Q2,Q3,Q4], and this vector is the voice feature vector used for BP-GA algorithm training, the wavelet decomposition result of speech signal is shown in Fig.2.

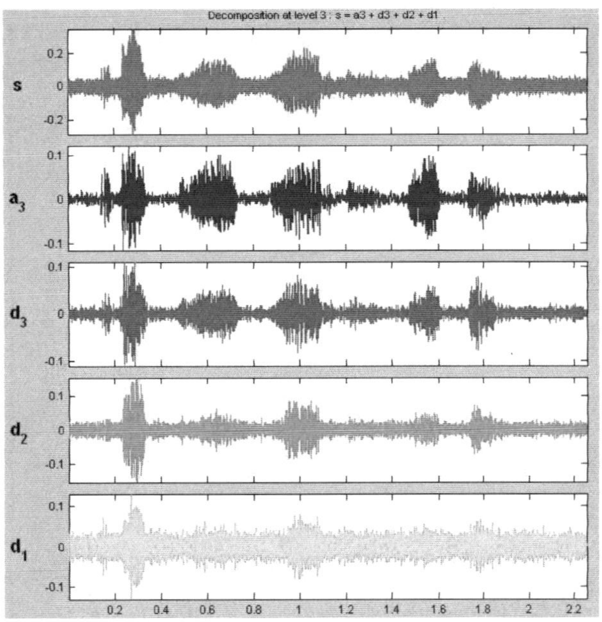

Figure 2. Wavelet decomposition results of password speech signal

3. Basic conception of BP-GA

BP-GA is a new optimization algorithm raised recently, it has the characteristics of high efficiency, strong global searching capability. GA is independent of gradient information, and has no requirement of the object function's continuity, even its definite expression [4]. BP-GA takes the advantage of artificial neural network and genetic algorithm, not only overcomes the low efficiency and long convergence time of GA in structure optimization, but also improves global solving capability of GA, so it is an effective and applicable solution[5-6]. Its working process is shown in Fig.3.

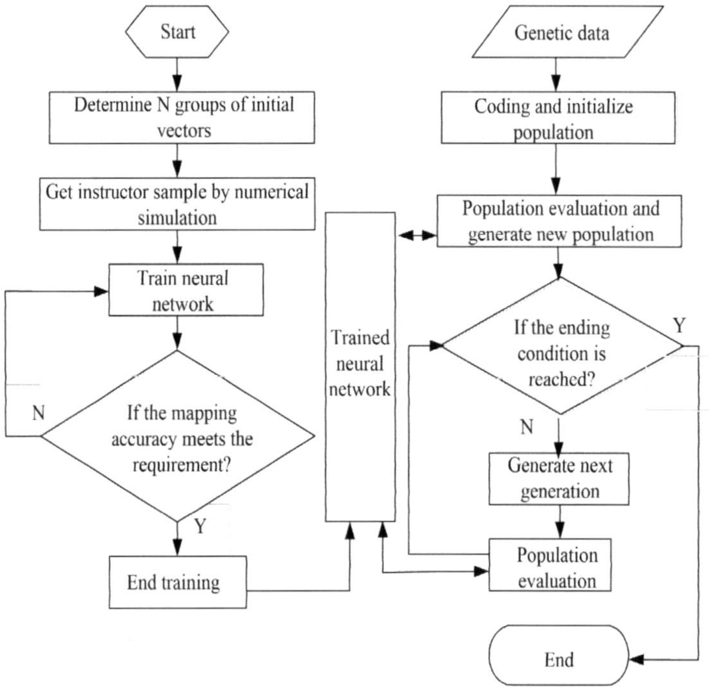

Figure 3. Flow chart of BP-GA

3.1 Construction of BP neural network voice print recognition

Multi-layer feed forward neural networks divides the network into several layers, which are arrayed in order, the neuron in layer i only receives signals from neuron in layer (i-1), and the neurons in all layers have no feedback. For a forward network, when vector x is inputted, vector y is outputted through the network, so the forward neural network can be seen as a transformer of x to y mapping[7]. The neural network's topological structure is shown in Fig.4.

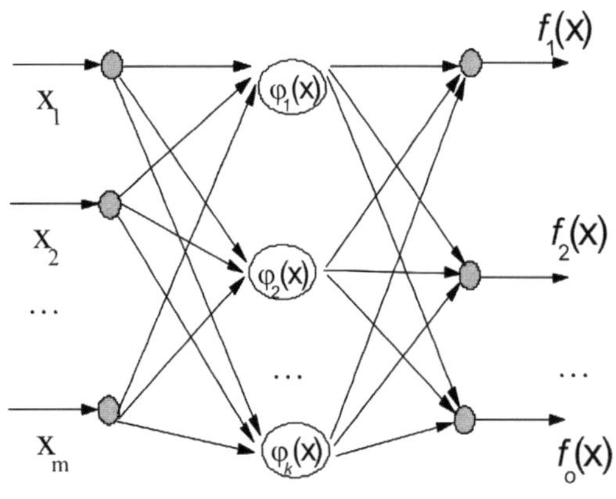

Figure 4. Topological Structure of BP Neural Network

Three layers BP neural network is employed. The input vector in the first layer can be adjusted according to practice. The first layer is a normalized layer, where three nodes are taken. The input vectors are software threat, hardware threat and management vulnerability. Experts are organized to form an evaluation group to make an evaluation of each factor. The values in $(0,1)$ are supposed as the risk evaluation results, i.e. the better the results are, the lower the risk of information subject's success is. The second layer is the input layer of BP network, five nodes are taken. The third layer is the output layer, one node is taken, the output characteristic function is s-style function, the output value is information construction risk degree, which are the continual numbers in $(0,1)$. $\omega 1_{ij}(i=1,2,3;\ j=1,2,3)$ is the weight of input layer, $o1_i(i=1,2,3)$ is the node of input layer, $\omega 2_j(j=1,2,3)$ is the weight of middle layer, $o2_i(i=1,2,3)$ is the node of middle layer, $\theta 2_j(j=1,2,3)$ is the weight of the second middle layer, θ_3 is the weight of output layer, ω_0 is risk degree. The mathematical expressions of the node are:

$$o2_j = \frac{2}{1+e^{-u2_j}} - 1$$

$$u2_j = \sum_{i=1}^{3} \omega 1_{ij} o1_i - \theta 2_j \qquad (j=1,2,3) \tag{4}$$

$$\omega = \frac{1}{1+e^{-u3}},\quad u3 = \sum_{j=1}^{3} \omega 2_j o2_j - \theta 3 \qquad (j=1,2,3) \tag{5}$$

3.2 Design of genetic algorithm

Genetic algorithm contains six basic parts: coding, initial population, fitness function, genetic manipulation and parameter control and end rules [8]. Coding is the bridge of problem and algorithm. In order to carry out genetic searching in a big space and improve algorithm accuracy, float-encoding is employed. Each individual of the initial population is composed of 40-bit binary strings, the value is determined as follows: generate a random number in $(0,1)$, if it is bigger than 0.5, the value of this bit is 1, or it is 0. Now the most used method is based on the penalty function idea. As the problem talked about in this paper is the least error optimization problem, then the fitness function can be expressed as:

$$Min\quad f = \frac{1}{\sum_{i=1}^{m} e_i^2},\ e_i = \omega_i - \overline{\omega}_i (i=1,2,\cdots,m) \tag{6}$$

Genetic manipulation contains five parts, they are selection, crossover, changing step, mutation and population update.

(1) Selection: individual is selected by roulette wheel method.

(2) Step setting: using scaling method

(3) Crossover: in an even crossover way, the mask sample is generated at random.

(4) Mutation: mutation bit is selected randomly, then get the bit value reversed.

(5) Population update: using pareto solution reserve strategy, the parent pareto solution is taken directly into child generation.

Two terminated rules are adopted here, for one rule, when the difference of the maximum and minimum of the object function is smaller than given precision $1e-6$, the algorithm is considered convergent, and the program is terminated. For the other rule, suppose the largest genetic generation value is 500, when the iteration time of algorithm reaches 500, the program is terminated. The operation results are shown in Fig.5.

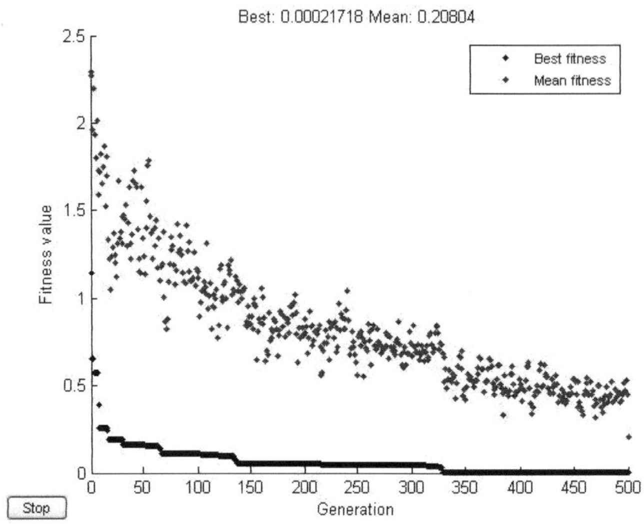

Figure 5. Results of Niche Genetic Algorithm

4. Simulation analyses of an example with BP-GA algorithm

Take 200 samples in speech database as feature vector in this example. The object output of real man is 1, and the false man is 0, the criterion for recognition is 0.5. make up MATLAB simulation procedure, the initial population is 500, crossover rate is 0.5, mutation rate is 0.042, the step scale is 0.8. take tansig function as the transfer function in the network's anonymous layer and satlin function as the transfer function in the output layer, which is single output to improve training speed. The Optimized training results of BP neural network is shown in Fig.6.

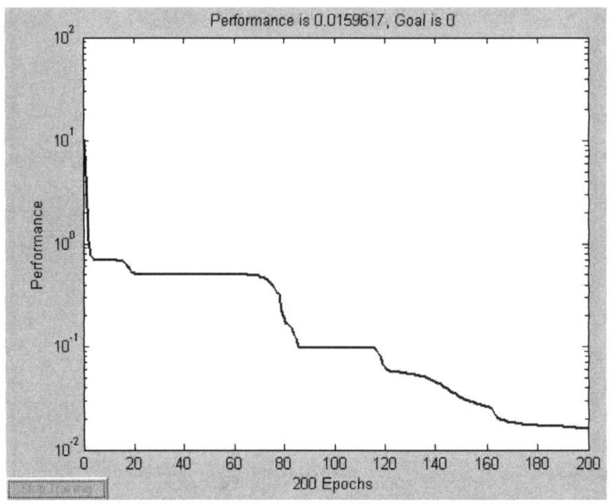

Figure 6. Optimized training results of BP neural network

From Fig.6, the false rate of recognition is about 1.6% using wavelet-BP-GA algorithm. To test the validity of the algorithm, we made a comparison with MFCC algorithm and LPCC algorithm, the detailed data of the experiment are shown in table.1.

Table 1. Comparison of wavelet-BP-GA with MFCC and LPCC

Recognition pattern	Training time/s	Recognition time/s	ognition rate Rec	False rate
MFCC	——	11.51	89%	5.1%
LPCC	——	12. 23	87%	4.4%
wavelet-BP-GA	325.86	9.16	95%	1.58%

5. Conclusions
Wavelet analysis can fully extract the signal's time-domain and frequency-domain features; BP network can make the recognition system perfect by gaining experience from fault; GA algorithm can improve the learning speed and convergence rate, decrease the network's false rate, and shorten the voice system's self-learning time. In short, compared with LPCC and MFCC, the wavelet-BP-GA pattern has the advantage of fast recognition velocity, high recognition rate, low fault rate, automatic error correcting, and strong robustness with different voice-makers.

6. References
[1] Xuelin Cao, Jinghua Li. Wavelet Packet Based Feature Extraction for Voice Recognition[J]. Computer Simulation, 2010, 27(11): 324-327.
[2] Qi Zhang. Shuxia Jiang, Xiangsheng li, etal. Spark Plug Gap Identification Based on Wavelet and Neural Network [J]. Computer Simulation, 2017, 34(4): 176-178.
[3] Koushan K. Automatic hull from optimization towards lower resistence and wash using artificial intelligence[C]. //Proceedings of FAST2003, 7th International Conference on Fast Seat Transportation. Ischia, Italy, 2003: 120-135.
[4] Zhonghua Li, Taishang Zhang. Research of fitness sharing niche genetic algorithms based on extension clustering[J]. Journal of Harbin Institute of Technology, 2016, 48 (5): 178-180.
[5] Yanfeng Kong, Zhenshan Zhang, Guangtao Cheng. Optimization Design of Elastic Constrained Mechanism in Swashplate Engine Based on Neural Network and Genetic Algorithm.[C]. 2010 International Conference on Computer, Mechatronics, Control and Electronic Engineering. Chang Chun: IEEE, 2010.
[6] Yanfeng Kong, Zhenshan Zhang, Guangtao Cheng. Optimization Design of Elastic Coupling in Swashplate Engine Based on BP-GA. [J]. Applied Mechanics And Mechanical Engineering, 2010. 32: 2380-2385.
[7] Shanhai Wang, Xinxing Jing, Haiyang Yang. Study of isolated speech recognition based on deep learning neural networks[J]. Application Research of Computers, 2015, 32 (8): 2289-2292.
[8] Deb K, Pratap A, Agarwal S, et al. A fast and elitist multi-objective genetic algorithm: NSGA-II [J]. IEEE Trans. On Evolutionary Computation, 2002, 6(2): 182-197.

IOP Conf. Series: Materials Science and Engineering **717** (2020) 012019 doi:10.1088/1757-899X/717/1/012019

The Research on Sensorless Control of Permanent Magnet Synchronous using SMC and SMO

Li Daoqi[1,2,3,4], Xiong Xinhong[1, *], Zhang Guoqing[1], Zhang Yanhui[2,3,4, *], Zuo Rui[1,2,3,4] and Feng Wei[2,3,4,*]

[1]School of Logistics Engineering, Wuhan University of Technology, Wuhan 430070, China
[2]Shenzhen Institutes of Advanced Technology, Chinese Academy of Sciences, Shenzhen 518055, China
[3]Guangdong Provincial Key Lab of Robotics and Intelligent System, Shenzhen Institutes of Advanced Technology, Chinese Academy of Sciences, Shenzhen 518055, China
[4]CAS Key Laboratory of Human-Machine-Intelligence Synergic Systems, Shenzhen Institutes of Advanced Technology, Shenzhen 518055, China

Feng Wei: wei.feng@siat.ac.cn; Xiong Xinhong: xiongxh@whut.edu.cn; Zhang Yanhui: zhangyh@siat.ac.cn

Abstract. This work proposes a way to achieve the sensorless control of permanent-magnet synchronous motor using sliding mode controller (SMC) and Sliding Mode Observer(SMO). In this Paper, the SMC is designed to replace the speed control loop PI controller from the original permanent magnet synchronous motor control system with the sliding mode variable structure control theory. Combined with the modified SMO, the "SMC and modified SMO" motor control system is proposed. Compared with the "PI and modified SMO" mode, the performance of the motor control is increased substantially, the interference torque speed attenuation is decreased by 87.5%, and the speed control error was also reduced by 67.7%. There is no overshoot with the reaction speed, and significant improvement in response speed, anti-interference performance, control accuracy and observation precision are achieved. Overall, both the system control and observation performance are significantly improved.

1. Introduction

The main disadvantage of the sliding mode observer is that the high frequency buffeting caused by the sliding mode variable structure control may increase the possibility of system crash, and it may also damage the system hardware. Therefore, the research on suppression of buffeting caused by sliding mode observer has become a research hotspot in recent years among researchers [1].

At present, in domestic areas, famous universities such like Harbin Institute of Technology, Hunan University are engaged in this program. Professor Li Tiecai's team from Harbin Institute of Technology has successively put forward adaptive sliding mode observer [2] and fourth-order hybrid sliding mode observer [3] of extended equation of state. The sliding surface of the two sliding mode observers is the difference between the actual current value and the estimated value, and through this to estimate the magnetic linkage of the motor.

In the field, because the PI controller is simple in structure, easy to set parameters, and it is easier to implement than other controllers, the PI controller is always used as the speed loop controller to form

IOP Conf. Series: Materials Science and Engineering **717** (2020) 012019 doi:10.1088/1757-899X/717/1/012019

a structure of "PI and SMO". However, there are still many serious defects existing, such as overshoot and poor robustness. In some high-precision control situations, it is very difficult for PI controllers to reach the requirements [4,5]. At present, the control of permanent magnet synchronous motor is developing into high precision high stability level. In order to obtain better control effect of the motor, it is necessary for us to replace the PI controller with a new type of controller. The sliding mode controller (SMC) has better robustness and speed than the PI controller, and the overshoot caused by PI control can also be avoided [6]. Nowadays, there are only a few literatures which have applied sliding mode observers ("SMC and SMO") to position sensorless control systems, thus it is very significant to study their specific effects.

2. SMC based on tan function and exponential approach law

The classic exponential approach law was proposed by Academician Gao Weibing [7], and its formula is as follows:

$$\dot{s} = -\varepsilon sign(s) - qs \tag{1}$$

In the middle ε, q is greater than zero constant. –qs is exponential approach, The q value affects the dynamic transition process of the control function. The larger the q value, the faster the point from the sliding surface reaches the sliding surface. The -εsign(s) is constant velocity trend, and the ε affects the rate at which the point reaches the sliding surface, and its rate decreases as the value of ε increases. By selecting an appropriate q value, not only the approach speed can be improved, but also the chattering can be effectively reduced.

The theoretical basis of the sliding mode controller and the sliding mode observer are the same, and they are all designed according to the sliding mode variable structure theory [8].

The motor voltage equation can be expressed as equation (2):

$$\begin{cases} u_d = R_s i_d + \frac{d}{dt}\psi_d - \omega_e \psi_q \\ u_q = R_s i_q + \frac{d}{dt}\psi_q - \omega_e \psi_d \end{cases} \tag{2}$$

ud、uq——Voltage component on d-axis and q-axis(V);

id、iq ——Current component on d-axis and q-axis(A);

ψd、ψq——Magnetic flux component on d-axis and q-axis(Wb).

Motor stator flux equation:

$$\begin{cases} \psi_d = L_d i_d + \psi_f \\ \psi_q = L_q i_q \end{cases} \tag{3}$$

The Ld is the d-axis inductance component and *Lq* is the q-axis inductance component(H).

Rewriting the mathematical model of the motor according to Equations 2 and 3.

$$\begin{cases} u_d = R_s i_d + L_s \frac{d}{dt} i_d - p_n \omega_m L_s i_q \\ u_q = R_s i_q + L_s \frac{d}{dt} i_q + p_n \omega_m L_s i_d + p_n \omega_m \psi_f \\ J\frac{d\omega_m}{dt} = \frac{3}{2} p_n \psi_f i_q - T_L \end{cases} \tag{4}$$

In the equation ω_m is the actual angular velocity of the motor(rad/s), TL is the load torque(N·m). Relationship between actual motor speed and electrical angular velocity: $\omega_e = \omega_m * p_n$.

Defining system state variables:

$$\begin{cases} x_1 = \omega_{ref} - \omega_m \\ x_2 = \dot{x}_1 = -\dot{\omega}_m \end{cases} \tag{5}$$

In the equation ω_{ref} is the set reference speed, according to Equation 2 and Equation 36:

$$\begin{cases} \dot{x}_1 = -\dot{\omega}_m = \frac{1}{J}(T_L - \frac{3}{2}p_n\psi_f i_q) \\ \dot{x}_2 = -\ddot{\omega}_m = -\frac{3}{2}p_n\psi_f i_q \end{cases} \tag{6}$$

Define the sliding surface as the $s=cx_1+x_2$, and deriving it

$$\frac{ds}{dt} = c\frac{dx_1}{dt} + \frac{dx_2}{dt} = cx_2 - \frac{3p_n\psi_f}{2J}\frac{di_q}{dt} \tag{7}$$

The equation of the sliding mode controller is as follows:

$$i_q = \frac{2J}{3p_n\psi_f}\int_0^t [cx_2 + \varepsilon sign(s) + qs]\, d\tau \tag{8}$$

Using the tanh function as the control function of the sliding mode controller, the improved sliding mode controller equation is as follows:

$$i_q = \frac{2J}{3p_n\psi_f}\int_0^t [cx_2 + \varepsilon\, tanh(s) + qs]\, d\tau \tag{9}$$

The structure block diagram of the sliding mode controller of Figure 1 can be constructed by Equation 9, and the difference between the actual speed and the set speed is taken as the input of the controller. The value is *x1*, and *x2* passing through the differential module to get *x1*. The current *iq* is used as the output of the sliding mode controller.

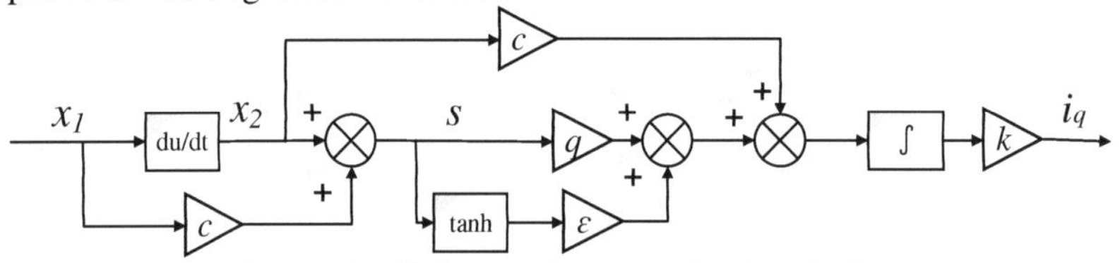

Figure 1. Sliding mode controller block diagram.

3. Simulation analysis of PMSM based on SMC and SMO

According to the mathematical model of permanent magnet synchronous motor, combined with vector control method and coordinate transformation relationship, the block diagram of the motor vector control system shown can be built and simulated by Matlab/Simulink. The speed loop and the current loop controller use the classic PI controller, and the position and speed signals are obtained by the motor position/speed detecting device. The speed loop PI controller parameters: kp = 0.15, ki = 7; the current loop PI controller parameters: kp = 9.35, ki = 3162.5. The sliding mode observer proposed by Zhang has a good observation performance [9]. This paper has established a modified SMO with reference to this document, which means removed the position/speed detection device and switched to the sliding mode observer module. Combined with the SMC mode that proposed in this paper, the PMSM control system of “SMC and SMO” is formed. The simulation structure block diagram is shown in Figure 2. Compared and analyzed with the traditional “PI and SMO” simulation results.

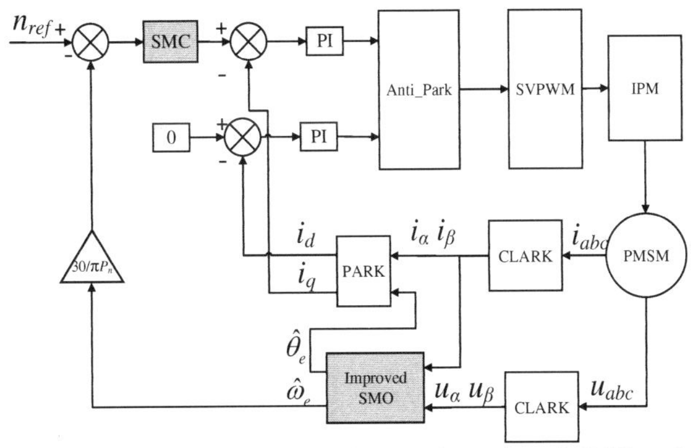

Figure 2. PMSM no position control block diagram based on “SMC and modified SMO”.

In order to obtain the anti-disturbance performance of the motor, the simulation is carried out under the disturbance of the speed switching and the disturbance of the load torque.

3.1 Simulation results under the influence of speed switching

The motor set has two speeds. The first speed is 300r/min, the second speed is 1000r/min, and the switching time is 0.06s.

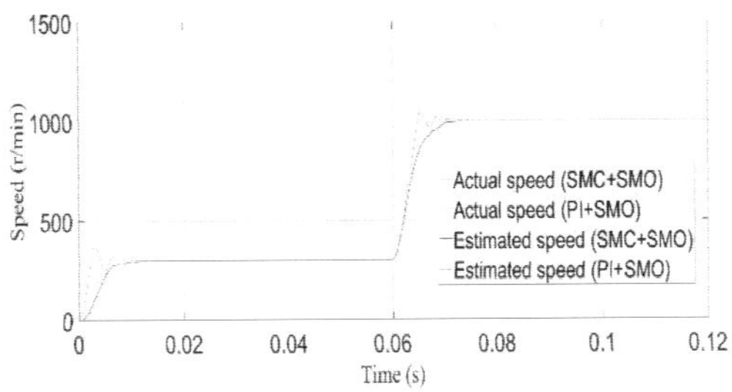

Figure 3. Speed response curve.

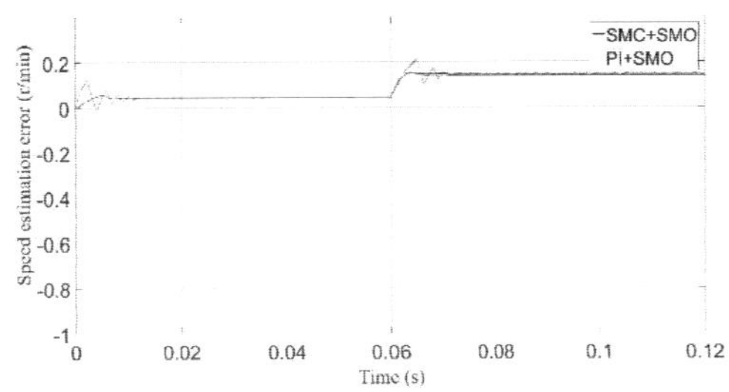

Figure 4. Speed error curve.

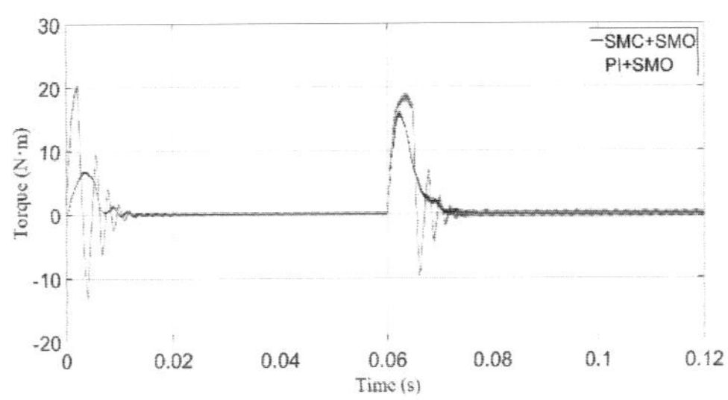

Figure 5. Torque response curve.

Figure 3 shows the speed response curve of the motor speed rising from zero speed to 300r/min and then accelerating to 1000r/min. The method of using SMC to improve SMO is faster than the “PI modified SMO” control, and the stability of the former’s speed switching is much higher than of the latter.

Figure 4 is the velocity observation error corresponding to Figure 3. Using the “SMC and modified SMO” method improve the accuracy of estimating the motor speed of the sliding mode observer in the starting phase and the speed switching phase.

Figure 5 shows the effect of speed switching on the torque of the motor. Under the “SMC and modified SMO” method, the torque has very good stability in both the starting phase and the speed switching phase.

3.2 Simulation results under torque interference

The reference speed set by the simulation is 1000r/min. When t=0.06s, a disturbance torque of 5N·m is applied to obtain the results of motor speed response curve, position curve, current curve, torque curve and observation error curve experiment.

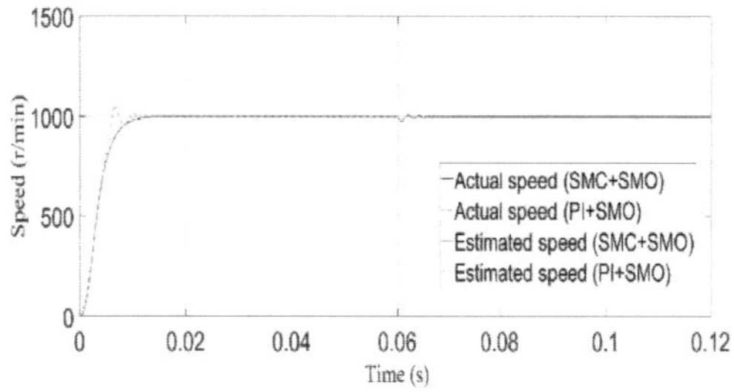

Figure 6. Speed response curve.

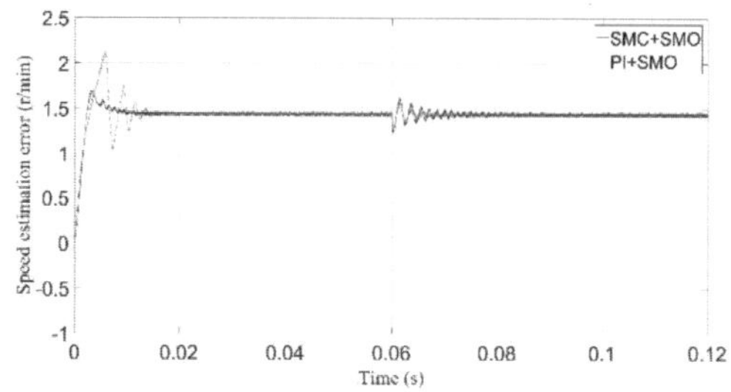

Figure 7. Speed error curve.

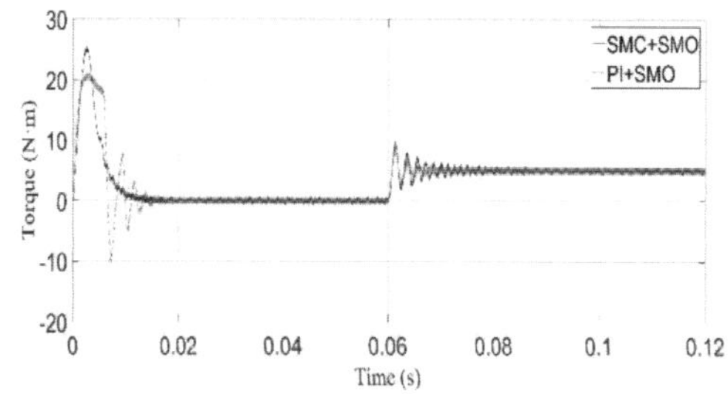

Figure 8. Torque response curve.

Figure 6 shows the speed response curve. The dotted line in the figure is the actual speed and estimated speed of the permanent magnet synchronous motor rotor under the "PI and modified SMO" method. The actual line is the actual speed and estimated speed of the permanent magnet synchronous motor rotor under the "SMC and modified SMO" method. It can be seen from the curve that under the "SMC and modified SMO" method, there is no speed overshoot and oscillation in the starting phase of the motor, and the start stability is obviously better than "PI and modified SMO".

Figure 7 shows the speed estimation error curve. After using the sliding mode controller, it not only improves the smoothness of the motor starting but also improves the speed tracking performance of the motor starting phase.

Figure 8 shows the torque response curve of the motor. The red line is the motor torque curve under the control of "PI and modified SMO" method, and the black line is the motor torque curve under the control of "SMC and modified SMO" method. It can be seen from the torque response curve that in the starting phase, the torque oscillation amplitude of the former is extremely large, which will reduce the life of the motor. Compared with the latter, the latter has better motor starting performance.

The specific evaluation indicators are as follows:

Table 1. System evaluation index.

Speed response evaluation index										Location response evaluation index
Control model	t_r (s)	t_p (s)	t_s (s)	σ_p (%)	ω_{re} (r/min)	N	K	S	Δω (r/min)	Δθ (rad)
PI and SMO	0.0061	0.0067	0.009	4.5	0.8~0.9	1	0.238	0.0008	0.17	0.000075
SMC and SMO	0.0053	/	0.008	0	0.2~0.3	0	0.238	0	0.14	0.000074

From Table 1, we can analyze the "SMC and modified SMO" method proposed in this paper. Compared with "PI and modified SMO", the rise time and adjustment time of the speed response are significantly improved, and the time is reduced by 16.7% and 10.2% respectively. No overshoot and oscillation, the actual motor speed error decreases from 0.8~0.9r/min to 0.2~0.3r/min, the speed control is more precise, and the stability is further improved.

4. Conclusions

In this paper, a "SMC and modified SMO" permanent magnet synchronous motor control system is established, which could enable the system possess a better capacity of resisting disturbance, significantly improves the reaction speed and the revolving speed control accuracy, eliminates the overshoot of the speed response.

5. References:

[1] Feng Hua-feng and Pan Hang-peng 2017 *A Study* of *SMO Chattering Elimination in Sensorless Control of PMSM* **30** 100-102.
[2] Su Jian-yong, Yang Gui-jie and Li Tie-cai 2008 *Extended state sliding mode observer of PMSM and estimation of rotor position plus speed* **05** 524-528.
[3] Su Jian-yong, Yang Gui-jie and Li Tie-cai 2009 *PMSM Sensorless Control Based on Four-order Hybrid Sliding Mode Observer* **29** 98-103.
[4] Yang M, Niu L and Xu D G 2013 *A novel piecewise Anti-Windup design for speed loop PI controller of PMSM servo system* 2013.
[5] Biswas P K 2016 *Design and Development of Closed Loop Model of an Adjustable Speed Permanent Magnet Synchronous Motor Drive Using PI Controller*
[6] Pan Guang and Wei Jing 2015 *Design of an adaptive sliding mode controller for synchronization of fractional-order chaotic systems* **64** 41-47.
[7] Gao Wei-bing and Cheng Mian 1989 *Quality Control of Variable Structure Control Systems* **4** 1-6
[8] LIU Jin-kun and SUN Fu-chun 2016 *Research and development on theory and algorithms of sliding mode control* **24** 407-418
[9] Lin S and Zhang W 2019 *Chattering reduced sliding mode control for a class of chaotic systems* **93** 1-10

Acknowledgements
The authors are grateful to the financial support from the National Natural Science Foundation of China (NO. U1813222), the National Natural Science Foundation of Guangdong (No.2016A030313177), Guangdong Frontier and Key Technological Innovation (No. 2017B090910013), the Science and Technology Innovation Commission of Shenzhen (No. JCYJ20170818153048647, No. JCYJ20180507182239617 and JSGG20170823140127645).

IOP Conf. Series: Materials Science and Engineering **717** (2020) 012020 doi:10.1088/1757-899X/717/1/012020

Acceleration Based Approach for Position Control

Nassim Khaled

Prince Mohammad Bin Fahd University, KSA

nkhaled@pmu.edu.sa

Abstract. In many current control applications, redundant sensors are becoming standard. In particular, autonomous cars, airplanes and aerospace applications all have a multitude of acceleration and position sensors. The need for robust control algorithms that can utilize multiple sensors, even for SISO systems, is fundamental, especially if these algorithms can accommodate sensor failures and account for system dynamics.
In this paper, we propose a combined position and acceleration feedback strategy to control position. The proposed strategy allows seamless embedding of the acceleration properties of the plant into the design of the controller.

1. Introduction

Autonomous machines, such as self-driving cars, have witnessed an accelerated growth the fields of sensor fusion, embedded controls and communication (Babak et al [1]). These sensors measure conditions outside the vehicle and vehicle position. Many of these sensors have overlapping and complementary capabilities [1]. Despite the improved capabilities of microcontrollers, vast majority of embedded control algorithms are still PID [2]. Significant body of research has been dedicated to advance PID controller tuning ([2], [3] and [4]) and structure [2]. In [5] and [6] they combine the readings from the inertial measurement unit and accelerometer sensor to control a flying quadrotor using PID. The structure of the PID remained the same. In this work, we propose a novel approach to embed the reading from two sensors into the design of the PID controller. Minimum and maximum accelerations can be used as lower and upper bound for the reference acceleration.

2. Structure of the Proposed Controller

In what follows we will present the structure of the multisensory based PID controller (Fig. 1). The position feedback is compared to the reference position to compute the position error. The position error will be transformed to acceleration reference via an acceleration function. This acceleration reference will be compared to an acceleration measurement to compute acceleration error. The acceleration error is the input to the PID controller.
The acceleration reference is designed to reduce the position tracking error. When the position measurement is less than the reference, the plant needs to be accelerated to reduce the position error. Similarly, when the position measurement is greater than the reference the plant needs to be decelerated. Introducing the acceleration function into the PID design allows us to include known dynamics such as maximum acceleration and deceleration values. It also allows us to shape the acceleration and deceleration profiles. The simplest function for acceleration is

$$\ddot{x}_{reference} = -k(x_{reference} - x_{measured}) \quad (1)$$

Where k is the range of acceleration divided by the range of position error.

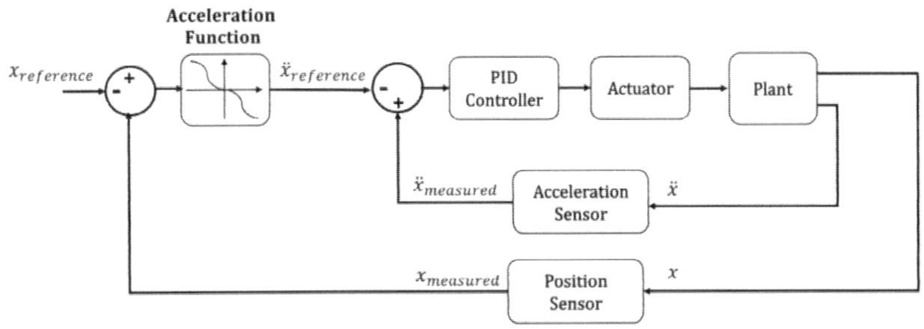

Figure 1. Block diagram of the proposed control strategy

Alternative functions for acceleration can accommodate physical constraints of the system such as minimum and maximum value of acceleration. This allows the designer to embed the physical constraints of the system in the design of the controller through shaping the reference. This is a valuable feature since linear PID controllers can't natively handle such constraints.

3. Application 1: Ball on Plate Control

To test the proposed controller, we use the ball on plate as a bench mark example. We bought the ball on plate hardware from roboholicmaniacs.com ([7]). It is a commonly used application for control concept testing ([8] and [9]). Acceleration measurement on this application was very noisy. The proposed controller is compared to the standard PID that got shipped with the hardware.

Another uniqueness about this benchmark system is its instability. It is a second order multi-input, multi-output problem (MIMO. The application has two inputs (two servo motors) and two outputs (x and y position of the ball). It has two servo motors to control the x and y slope of the plates. The sensing mechanism of the ball is a four-wire resistive touch screen (Fig. 2). The microcontroller used is ATmega2560 Microchip and it is compatible with Arduino software toolset.

The ball is 1 inch in diameter and its mass is 51grams. Four wire resistive touch screens are noisy, even in the presence of a consistent force applied on it. It needs significant filtering effort to get a clean signal that represents the position of the ball. The standard Matlab® linear Kalman filter block was used to produce acceptable position and acceleration measurements.

The servos used are RC servo 25KG with metal gear. Their speed at zero load is 0.15seconds/60°. The servos can be commanded from 00 to 180°. For the ball on plate application with the current plate and arm dimensions, we limited the command of the servos to the range [70°,110°].

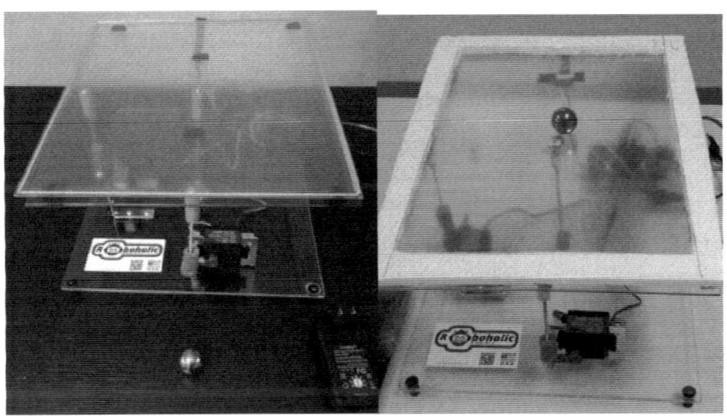

Figure 2. Original and modified ball on plate setup

Fig. 3 shows the raw measurement for one of the experiments sampled at 0.1 seconds. The resistive touch screen measurement exhibits noise due to two main factors. The first being the four wires connecting the screen to the controller are subjected to vibrations due to the real time control of the ball. The second is the fast response time of the servos which causes the ball in some cases to lose contact with the plate or, in less extreme cases, have a smaller action force on the plate.

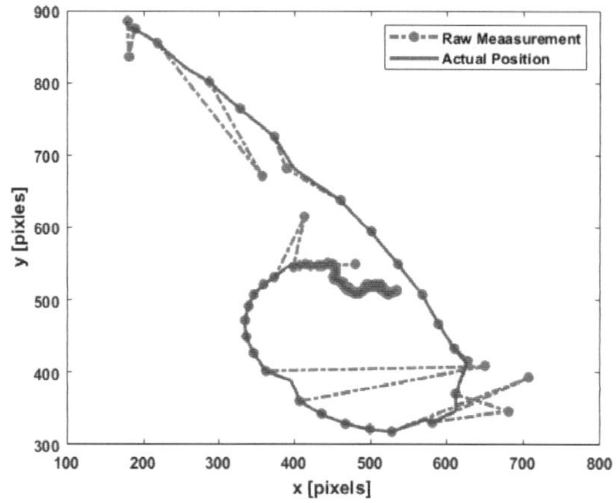

Figure 3. Raw and actual position of the ball

To provide the controller with good measurements, the position of the ball was measured every 0.01 seconds. Two standard Matlab® SISO Kalman filters for x and y measurements were deployed.

The objective of the ball on plate controller is to keep the ball at a specific location on the plate. In our case, we are trying to maintain the ball at the location [512,512] which is the center of the plate. The touch screen provides the raw measurement of the ball in the x and y directions. These measurements are input into two linear Kalman filters. The output of the filter is compared to the reference, which is 512, to compute the error in the x and y directions. The error is fed to two PID controllers which command two servos (Fig. 4).

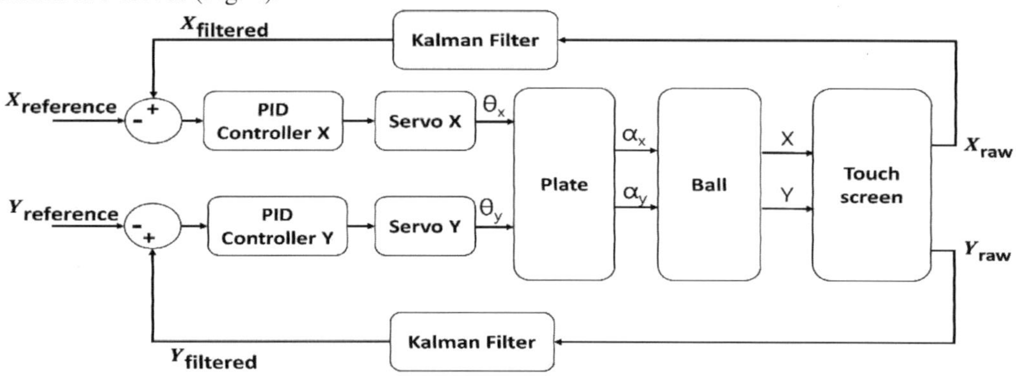

Figure 4. PID position control

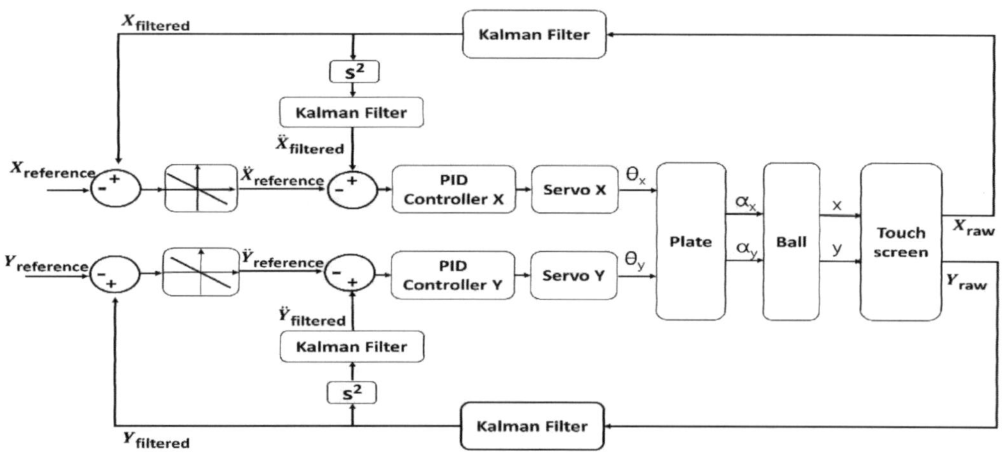

Figure 5. Modified PID position control

IOP Conf. Series: Materials Science and Engineering **717** (2020) 012020 doi:10.1088/1757-899X/717/1/012020

The modified PID controller is applied on the ball on plate to assess its feasibility. Fig. 5 shows the block diagram for the controller.

Table 1. Parameters for the PID Controller

Tuning Parameters	Traditional PID	Modified PID
Proportional	0.05	0.03
Integral	0.0001	0.0001
Derivative	0.4	0.4
Execution time (s)	0.1	0.1

Fig. 6 shows the position control results for the two controllers. The modified controller got the ball closer to the target with minimal overshoot. Controlling the ball to hit the reference point directly turned to be a challenging task given the nonlinearity of the system due to friction. Traditional PID had an overshoot that was challenging to eliminate. This can be clearly seen in Fig. 7.a. In the same figure, we can see that the modified PID response very fast in the first approach (the first 3 seconds). From there on, past the 3 seconds mark, the control action is slowly maturing under the impact of the integral term. The integral term works on disturbances due to friction and ball getting stuck on the plate as well as non-zero sloped surfaces where the plate is placed. Fig. 7.b shows the servo command comparison between the two controllers. There is clear chattering in the performance of the traditional during several instances of the simulation. In particular, we highlighted the region around the 50 seconds mark. On the other hand, the modified PID was much steadier when the position error was relatively small.

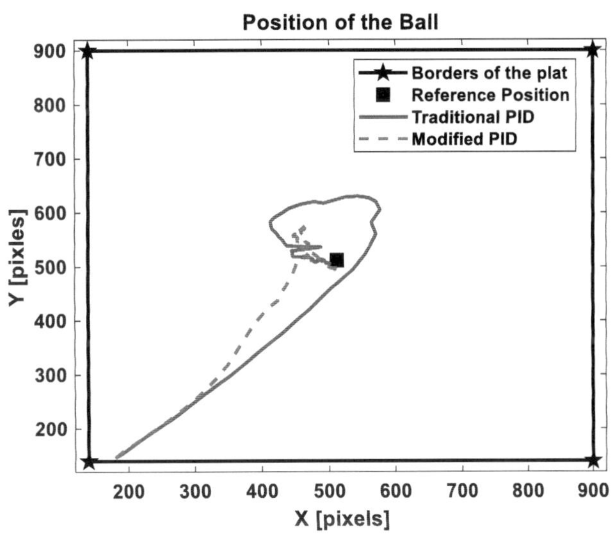

Figure 6. Raw and actual position of the ball

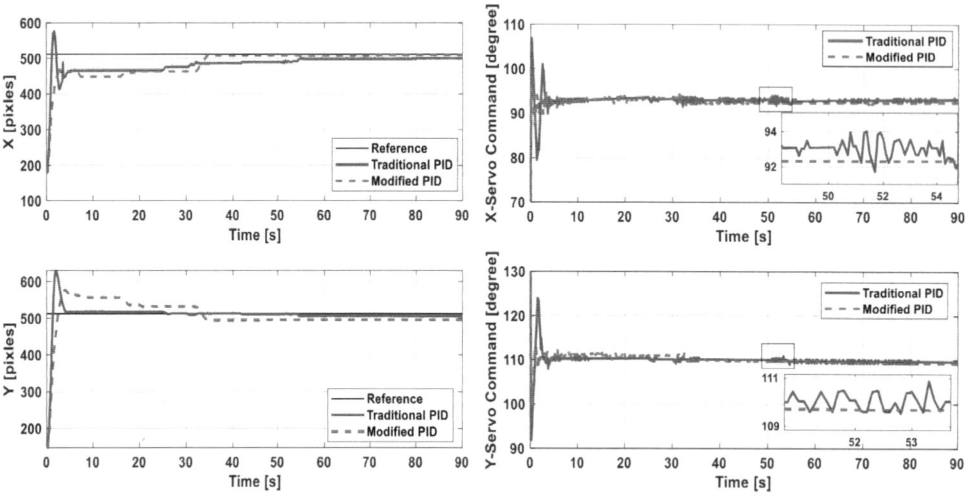

Figure 7. Tracking performance and servo command comparison for both controller

4. Conclusions and Future Work

In this paper we proposed a modified PID controller that allows utilizes the acceleration and position measurements to control the position. The proposed strategy was experimentally tested on the ball on plate system. Despite the noisy position and acceleration measurements, the performance of the modified controller was better than that of the traditional PID controller. The tuning effort was comparable to that of the traditional PID. Future work will include theoretical analysis for robustness to study impact of acceleration reference.

5. References

[1] Babak, S. J., Hussain S. A., Karakas B. and Cetin S., “Control of autonomous ground vehicles: a brief technical review,” IOP Conference Series: Materials Science and Engineering, vol. 224, con. 1, 2017.

[2] Antonio Visioli, “Research trends for PID controllers,” 2012, Acta Polytechnica 52(5):144-150

[3] Kaur, M., "Literature Review of PID Controller based on Various Soft Computing Techniques,"2017, IJRITCC, 5(6): 1375 – 1378

[4] Vaishnavi, P. and Balasubramanian, G. "Tuning of PID Controller for Cascade Unstable systems Using Genetic Algorithm," 2018, International Journal of Pure and Applied Mathematics, 118(18), 2011-2019.

[5] Zul Azfar, A. and Hazry, D., 2011 “A simple approach on implementing IMU sensor fusion in PID controller for stabilizing quadrotor flight control," IEEE 7th International Colloquium on Signal Processing and Its Applications, Penang, Malaysia

[6] Noordin, A., Ariffanan, M., Basri, M. and Mohamed, Z., 2018, "Sensor fusion for attitude estimation and PID control of quadrotor UAV," International Journal of Electrical and Electronic Engineering and Telecommunications, 7, (4)

[7] www.roboholicmaniacs.com

[8] Debono, D. and Bugeja, M. "Application of Sliding Mode Control to the Ball and Plate Problem," 2015, 12th International Conference on Informatics in Control, Automation and Robotics (ICINCO-2015), 412-419

[9] Nokhbeh, M. and Khashabi, D., "Modelling and Control of Ball-Plate System," 2011, Ph.D. thesis Amirkabir University of Technology.

Acknowledgments

The author wishes to acknowledge the support from Prince Mohammad Bin Fahd University.

IOP Conf. Series: Materials Science and Engineering **717** (2020) 012021 doi:10.1088/1757-899X/717/1/012021

Design of PI Controller for PMSM using Chaos Particle Swarm Optimization Algorithm

Zuo Rui[1,2,3,4], Xiong Xinhong[1,*], Chen Lianbo[1], Guo Shifeng[2,3,4,*], Zhang Yanhui[2,3,4] Li Daoqi[1,2,3,4] and Feng Wei[2,3,4,*]

[1]School of Logistics Engineering, Wuhan University of Technology, Wuhan 430070, China
[2]Shenzhen Institutes of Advanced Technology, Chinese Academy of Sciences, Shenzhen 518055, China
[3]Guangdong Provincial Key Lab of Robotics and Intelligent System, Shenzhen Institutes of Advanced Technology, Chinese Academy of Sciences, Shenzhen 518055, China
[4]CAS Key Laboratory of Human-Machine-Intelligence Synergic Systems, Shenzhen Institutes of Advanced Technology, Shenzhen 518055, China

Corresponding author's E-mail contact: wei.feng@siat.ac.cn; xiongxh@whut.edu.cn; sf.guo@siat.ac.cn

Abstract. The proportional-integral-derivative (PID) controller parameters tuning is critical for improving the performance of permanent magnet synchronous motor (PMSM) systems. To improve the performance of particle swarm optimization (PSO) in the PID parameters tuning, the chaos particle swarm optimization (CPSO) is adopted to tune PI parameters in this work. The PMSM vector control system is selected as the research object, and its speed loop PI controller using CPSO algorithm is implemented. The optimal PI controller parameters are obtained using the CPSO algorithm with ten iterations. The simulation is conducted using MATLAB/Simulink to evaluate to the PI controller performance. Four performance indicators including overshoot, rise time, peak time and ITAE are used to evaluate the PI controller with CPSO, and the results suggest that the CPSO algorithm has better performance and stability compared with other tuning methods.

1. Introduction

The permanent magnet synchronous motor (PMSM) has the advantages of high efficiency, high power density, high torque inertia ratio, low noise, and maintenance-free [1], and has been widely used in electric vehicles, aerospace and robotics etc. Its control research is therefore receiving increasing attention [2]. Since its introduction, the PID controller has been widely used in various fields of modern industry because of its simple structure and excellent performance. Considering the performance of the PID controller is closely related to the PID parameters [3], it is therefore critical to study the tuning technique of PID parameters.

With the development of swarm intelligence algorithm, researchers have used it for parameter tuning of PID controllers. Ouyang [4] used differential evolution (DE), genetic algorithm (GA) and PSO to optimize the gain of the position domain of PID controller to improve the contour tracking performance of robot manipulator. Wang [5] proposed a hydro-generator excitation controller based on PSO and fuzzy PID. The initial parameters of the controller were determined by the PSO and the system was dynamically controlled by the FAPID. In addition, compared with the traditional PID and fuzzy PID control, the simulation results demonstrate that the fuzzy adaptive PID based on PSO algorithm had faster response speed and smaller overshoot. Jaafar [6] proposed a PID control method for a high nonlinear double pendulum bridge crane without payload motion feedback signal. Using the improved PSO algorithm, the optimal parameters of the PID controller were achieved according to the vertical distance oscillation and potential energy of the crane. Compared with the conventionally adopted PSO algorithm based on horizontal distance, this method achieved high performance for the less complicated controller.

In the evolutionary computational method, the PSO algorithm has robustness in solving continuous nonlinear optimization problems [7]. Compared with other methods, PSO method can produce high-quality solutions with less computational time and has stable convergence characteristics. However, the initial particle quality of the PSO algorithm is unstable, and it is easy to fall into local extremum during the iterative process [8]. In order to solve the shortcoming of PSO, a modified PSO algorithm based on chaos theory for PID tuning based on PMSM is proposed in this work. The tuning effect of the modified PSO algorithm on the parameters of the PI controller has been analyzed with the MATLAB/Simulink and the result indicates that this algorithm is effective for tuning the PI controller parameters of the PMSM servo system.

2. Mathematical model and methods

2.1. Mathematical Model of PMSM

The mathematical model of PMSM in rotor reference frame can be expressed as follows [9]:

$$u_d = Ri_d + L_d \frac{d}{dt} i_d - \omega_e L_q i_q \tag{1}$$

$$u_q = Ri_q + L_q \frac{d}{dt} i_q + \omega_e \left(L_d i_d + \psi_f \right) \tag{2}$$

$$\frac{d}{dt} \omega_m = \frac{1}{J} \left(T_e - T_L - B\omega_m \right) \tag{3}$$

Where, u_d and u_q are the voltage in the q-axe and d-axe; i_d and i_q are the q-axe and d-axe current; R is stator resistance per phase; ψ_f is armature flux linkage due to rotor magnets; L_d and L_q are quadrature and direct axis stator self-inductance in rotor reference frames; Te is the electromagnetic torque; T_L is the Load Torque; ω_e is electrical rotor speed; ω_m is mechanical rotor speed.

In the surface-mount PMSM, the stator self-inductance satisfies $L_d=L_q=L_s$, and therefore the electromagnetic torque equation is:

$$T_e = \frac{3}{2} p_n i_q \psi_f \tag{4}$$

2.2. Chaos particle swarm optimization algorithm

In a D dimensional search space, the position of particle i is represented as $x_i^t=(x_{i1}^t,x_{i2}^t,\ldots,x_{iD}^t)^T$, and the velocity of particle i is represented as $v_i^t=(v_{i1}^t,v_{i2}^t,\ldots,v_{iD}^t)^T$. The best previous fitness and corresponding position of particle i is reflected as P_{best} and $P_i^t=(P_{i1}^t,P_{i2}^t,\ldots,P_{iD}^t)^T$, respectively. Fitness and position of the best particle in the swarm is represented as g_{best} and $P_g^t=(P_{g1}^t,P_{g2}^t,\ldots,P_{gD}^t)^T$. In the standard PSO, the search process can be express as [10]:

$$v_{id}^t = wv_{id}^{t-1} + c_1 r_1 \left(p_{id}^{t-1} - x_{id}^{t-1} \right) + c_2 r_2 \left(p_{gd}^{t-1} - x_{id}^{t-1} \right) \tag{5}$$

$$x_{id}^t = x_{id}^{t-1} + v_{id}^t \tag{6}$$

Where, t is the iteration number; w is the inertia weight; r_1 and r_2 are uniformly distributed random numbers with range [0,1]; c_1 and c_2 are called cognitive and social acceleration coefficients respectively.

In the PSO algorithm, an evaluation function is selected to calculate the quality of particles' position. The Integrated time and absolute error (ITAE) are used as the fitness function in the paper, which is defined as:

$$F = \int_0^t t\left|e(t)\right|d(t) \tag{7}$$

Logistic equation is a typical chaotic system, which is expressed as [11]:

$$z(t+1) = \mu z(t)[1 - z(t)] \tag{8}$$

Where, $z \in [0,1](z(0) \neq 0, 0.25, 0.50, 0.75, 1)$, μ is the control parameter and usually set to 4.

In the process of particle swarm initialization, chaos initialization can generate a large number of ordered chaotic particles and improve the quality of initial particles. Aiming at the problem that particles are easily trapped in local extremum in PSO algorithm, chaos disturbance is added in the search process to make particle jump out of local extremum and realize global optimization [12].

Chaos theory mainly includes chaos initialization and chaos disturbance. The process of CPSO is as follows:

(1) Initialize the particle's position and velocity with the chaos, and the initial groups are selected according to the principle of fitness-best selection.

(2) Add a disturbance Δx to its existing position of each particle to produce a new particle. Calculate the fitness value of the new and old particle, and select the particle with better performance to form a new particle group.

(3) Update the individual extremum and the global extremum to obtain corresponding particle.

(4) Update the particle's position and velocity according to expression (5) and (6).

(5) Detect convergence condition. If the termination condition is satisfied, the iteration is terminated and the optimal solution is output. Otherwise, return (2).

2.3. Design of Speed Loop PI Controller Based on CPSO Algorithm

PMSM has speed control, torque control and position control. Speed control is the most common control method in servo systems. The performance of the system is determined by the speed loop. Therefore, CPSO algorithm is adopted to set the PI parameters of the speed loop in the paper. The process of tuning the speed loop PI parameters using CPSO algorithm is shown in Figure 1.

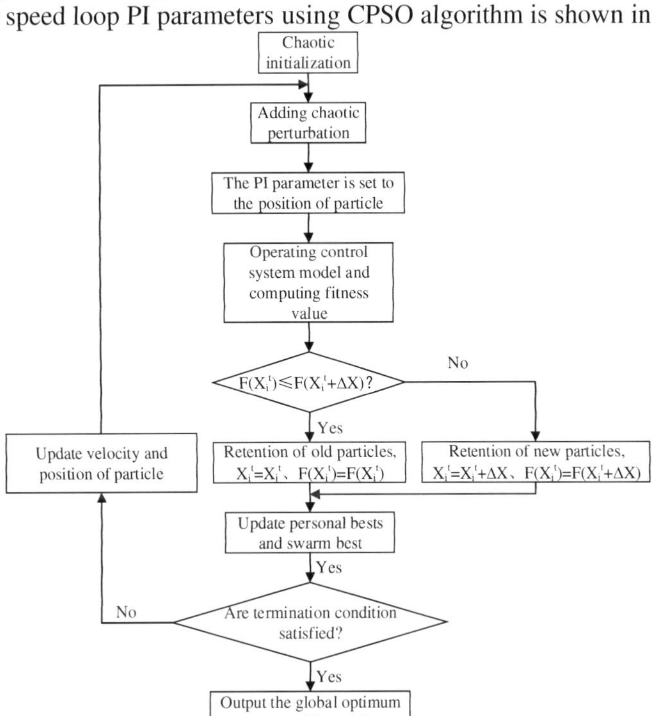

Figure 1. The process of tuning the PI parameters using CPSO algorithm.

IOP Conf. Series: Materials Science and Engineering **717** (2020) 012021 doi:10.1088/1757-899X/717/1/012021

3. Results and discussion

According to the PMSM vector control and the optimization principle of CPSO algorithm, the PMSM vector control schematic based on CPSO is implemented as follows.

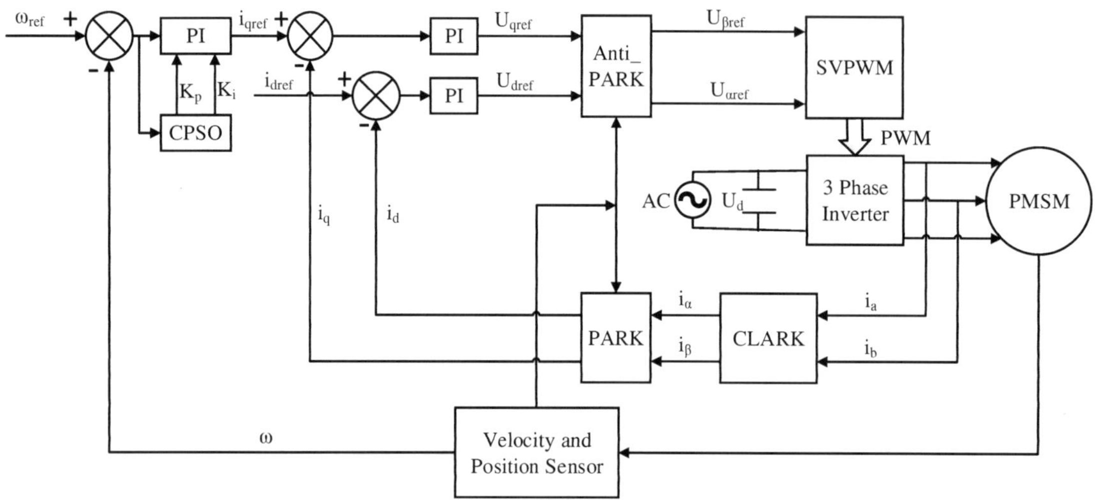

Figure 2. PMSM vector control schematic based on CPSO.

The number of the chaotic particles and iterations are set at 100 and 30, and the reference speed is predetermined at 1000 r/min. According to the parameter optimization process of PI controller using CPSO, the fitness curve is shown in Figure 3(a), and the optimization curve of proportional coefficient K_p and integral coefficient K_i is shown in Figure 3(b) and (c), respectively.

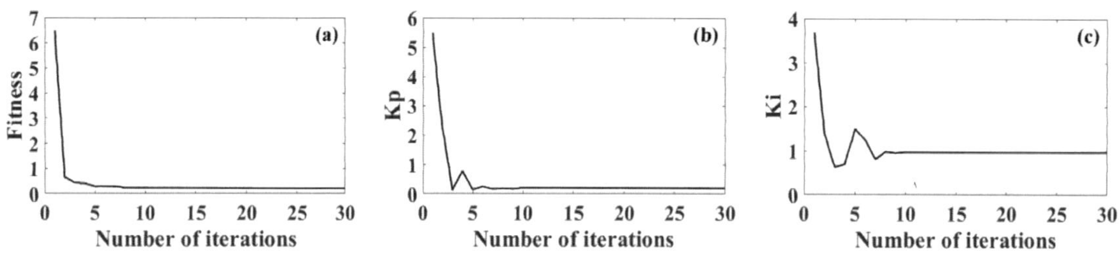

Figure 3. The process of optimizing PI controller parameters using CPSO algorithm.

The results suggests that the system obtains the best parameter at the 10th iteration, before reaching the maximum number of iterations. As shown in Figure 3(b) and (c), the parameters set by CPSO algorithm are K_p=0.21 and K_i=0.97.

In order to verify the performance of the PI CPSO algorithm, the performance of the PMSM system using different tuning methods is compared. The traditional PI controller has been manually debugged with the parameters of K_p = 0.14 and K_i = 7. After optimization by PSO algorithm, the PI controller parameters are determined at K_p= 0.18 and K_i= 3.95. The parameters of the PI controller obtained by the CPSO algorithm are K_p=0.21 and K_i=0.97, respectively. The speed response curve corresponding to two different parameters of the PI controller in two cases is shown in Figure 4.

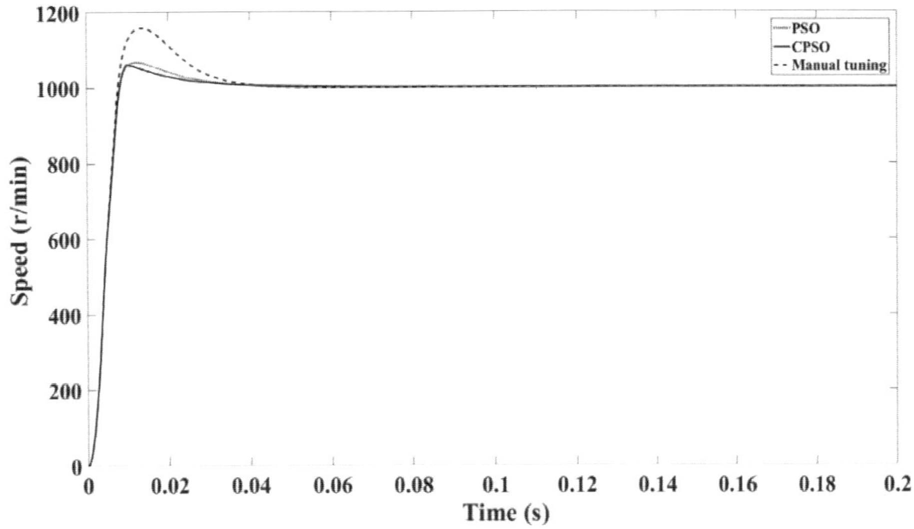

Figure 4. PMSM system speed response curve.

As shown in Figure 4, it is difficult to obtain satisfactory results by manual tuning, and the amount of overshoot is large. The PI controller parameters obtained by PSO algorithm have less overshoot than manual tuning, and the system based on PSO algorithm has faster response speed. Compared with PSO algorithm, CPSO algorithm has improved substantially in response to speed and overshoot indicating that the CPSO algorithm is effective for tuning the PI controller parameters of the PMSM servo system.

In order to quantify the performance difference of the three methods, the CPSO algorithm, the PSO algorithm and the manual tuning are compared from the following aspects, overshoot (σ_p), rise time (t_r), peak time (t_p), and integral of error and time (ITEA) (Table 1).

Table 1. PMSM performance parameters.

Tuning method	σ_p(%)	t_r(s)	t_p(s)	ITEA
Manual tuning	15.81	7.65×10^{-3}	13.54×10^{-3}	0.0584
PSO	6.62	8.09×10^{-3}	11.89×10^{-3}	0.0334
CPSO	6.10	8.11×10^{-3}	9.90×10^{-3}	0.0306

4. Conclusion

In this paper, the CPSO, a modified PSO algorithm based on chaos theory, has been developed to mitigate premature convergence problem of conventional PSO. The CPSO is applied to tuning the PI controller parameter for tuning the PMSM system. For validating CPSO, its performance is compared with conventional PSO and manual tuning method. The simulation results suggest that the CPSO algorithm can improve the tuning efficiency. In addition, the PMSM servo system based on CPSO algorithm has achieved better performance.

5. Reference

[1] Kommuri S K, Defoort M, Karimi H R and Veluvolu K C 2016 *IEEE Transactions on Industrial Electronics* **63** 7671-81
[2] Zhang X G, Hou B S and Mei Y 2017 *IEEE Transactions on Industrial Electronics* **32** 3818-34
[3] Shah P and Agashe S 2016 *Mechatronics* **38** 29-41
[4] Ouyang P and Pano V 2015 *Algorithms* **8** 697-711

[5] Wang Z, Zhang L and Peng Z 2017 *Hunan Daxue Xuebao/Journal of Hunan University Natural Sciences* **44** 106-111
[6] Jaafar H I, Mohamed Z, Mohd S and Husain A R 2019 *JVC/Journal of Vibration and Control* **25** 907-921
[7] Zhang J K 2009 *Computer Engineering and Application* **45** 43-45
[8] Jordehi A R 2015 *Applied Soft Computing* **26** 401-417
[9] Niu F, Wang BS, Babel AS and Li K 2016 *IEEE Transactions on Power Electronics* **31** 1408-24
[10] Jordehi and Rezaee A 2015 *Applied Soft Computing* **26** 401-417
[11] Tan D 2012 *Lecture Notes in Electrical Engineering* **144** 469-476
[12] Liu J and Gao Y 2008 *Journal of Computer Applications* **28** 322-325

Acknowledgements
The authors are grateful to the financial support from the National Natural Science Foundation of China (NO. U1813222), the National Natural Science Foundation of Guangdong (No.2016A030313177), Guangdong Frontier and Key Technological Innovation (No. 2017B090910013), the Science and Technology Innovation Commission of Shenzhen (No. JCYJ20170818153048647 and No. JCYJ20180507182239617).

IOP Conf. Series: Materials Science and Engineering **717** (2020) 012022 doi:10.1088/1757-899X/717/1/012022

Numerical Investigation of Rough Micro- and Mini-channel Heat Sinks for Varying Aspect ratio

Munib Qasim Ansari *, Guobing Zhou

School of Energy, Power, and Mechanical Engineering, North China Electric Power University, Beijing 102206, PR China

*E-mail address: munibqasim@icloud.com

Abstract. Numerical simulations are performed to study the impact of aspect ratio on the micro- and mini-channel heat sink behaviours in the presence of sinusoidal surface roughness. A CFD model is first validated with the relevant reference data for laminar flow conditions using ANSYS-Fluent analysis. The found results of this validation study show that the simulation results agree well with the reference data. The considered channel height, absolute roughness height and diameter are 250 μm, 30 μm and 366 to 374 μm, respectively. The top and bottom rough walls are under constant heat flux conditions, while other walls are adiabatic with air as the working fluid. The micro- and mini-channels with long wetted perimeter and lower aspect ratio have shown more convective heat transfer but with slightly higher frictional resistance. The convective heat transfer and thermal performance factor both improve with the increase in Reynolds number where all the values are higher than 1. The maximum value of the thermal performance factor is noted up to 1.6 for the channel having aspect ratio of 0.038 (Reynolds number = 250). This study shows that micro- and mini-channel heat sinks are sensitive to surface roughness, Reynolds number where varying aspect ratio have influence on both flow and heat transfer performances in laminar regime.

1. Introduction

The heat dissipation of electronic devices has become an important factor constraining the rapid development of them. Nowadays, electronic devices tend to be highly integrated within a smaller volume. In this regard, a lot of work has been done to improve the performance of heat sinks without compromising the compactness [1]. Qu and Mudawar [2], and Xi et al. [3] investigated micro- and mini-channels extensively by performing different experiments. In most of these studies, several discrepancies have been reported between the macro-scale correlations and micro-scale flows [4]. The different ways were also applied to enhance the heat transfer performance of micro-channel heat sinks by varying channel shapes [5], heating methods [6] and surface roughness [7].

The configuration of surface roughness can increase the heat transfer performance but at the expense of high-pressure drops. According to Kharati-Koopaee and Zare [8], structured surface roughness is recommended for numerical modeling instead of random surface roughness due to the simplicity of modeling and more detail insight into the flows. Zhang et al. [9] numerically studied the effects of 2D surface roughness and reported that semi-circular shaped roughness could give better performance with fewer flow circulations. Dharaiya et al. [7] also numerically investigated the effects of sinusoidal roughness and found that the maximum heat transfer performance with the maximum deviation from smooth channels was 6.33. Ghani et al. [10] studied the effects of ribs on micro-channels and reported that the maximum performance factor of 1.85 (relative cavity amplitude = 0.15, relative rib width = 0.3 and relative rib length = 0.5, and Reynolds number = 800). Guo et al. [11] also numerically studied the effect of surface roughness and concluded that 3D models are more accurate.

There is not sufficient literature for air as working fluid [8]. Therefore, in this work, micro- and mini-channels are structured with sinusoidal surface roughness and investigated numerically with varying channel width and air as working fluid.

2. Channel and roughness configurations

A single rectangular micro- and mini-channel heat sink is considered for simulation. The top and bottom walls of the channel are devised with sinusoidal surface roughness, as shown in Fig. 1. The width, height, and length of the channel are represented as '*W*',' *H*', and '*L*,' respectively. The working fluid-air flows along the channel length (*x*-axis). Considering the coordinate system shown in Fig. 1, the bottom and top walls are obtained by Eqs. (1) and (2), respectively [8]. Fig. 1(b) elaborates further details where surface roughness pitch, height, and constricted channel height are represented by *s*, *r*, and H_c, respectively.

$$y(x) = r\cos^4\left(\frac{\pi}{s}x\right) \quad \text{Bottom wall} \quad (1) \qquad y(x) = H - r\sin^4\left(\frac{\pi}{s}x\right) \quad \text{Top wall} \quad (2)$$

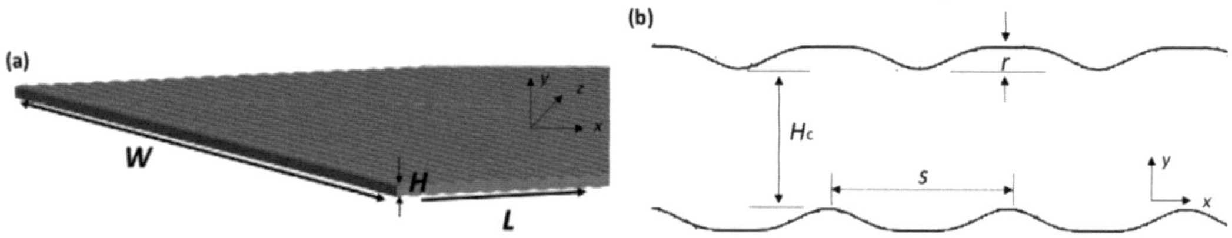

Fig. 1. Schematic of rough micro- and mini-channel heat sink: (a) Stereogram; (b) Front view.

The constricted flow parameter scheme defined by kandlikar et al. [7] is considered for modeling. The corresponding constricted channel height (H_c), hydraulic diameter (D_c) and aspect ratio (α_c) are defined as,

$$H_c = H - 2r \quad (3) \qquad D_c = \frac{4H_cW}{2(H_c+W)} \quad (4) \qquad \alpha_c = {H_c}/{W} \quad (5)$$

The channel length, channel height, surface roughness height, and pitch are considered as 12.5 mm, 250 μm, 30μm, and 250 μm, respectively, for all the cases. The aspect ratio variations are based on channel widths (*W*) of 12.7 mm, 8.95 mm and 5 mm, and respective values of constricted aspect ratios (α_c) are 0.015, 0.021 and 0.038.

3. Numerical method

3.1. Governing equations and assessment parameters

The simulated flow is 3D, steady, and assumed incompressible with constant properties. The radiation and gravitational force are neglected in this work. The heated walls are under constant heat flux, while thermo-physical properties are assumed to be constant. The assumptions of no-slip boundary condition, the flow continuum, and Navier-Stokes equation are reasonable in laminar flow regime of the study. The governing equations of continuity, momentum, and energy are written based on the above assumptions:

Continuity equation: $\nabla\vec{V} = 0$ (6)

Momentum equation: $\rho(\vec{V}.\nabla\vec{V}) = -\nabla P + \nabla.(\mu\nabla\vec{V})$ (7)

Energy equation: $\rho C_P(\vec{V}.\nabla T_f) = k_f\nabla^2 T_f$ (8)

where *P*, C_p, μ, T_f, and k_f are fluid pressure, specific heat, dynamic viscosity, fluid temperature, and thermal conductivity, respectively.

For the simulations results, constricted Reynolds number (Re_c), Fanning friction factor (f_c), and Nusselt number (Nu_c) are calculated (using constricted parameters) and defined as

$$Re_c = \frac{\rho U d_c}{\mu} \quad (9) \qquad f_c = \frac{\Delta P d_c}{2\rho U^2 L} \quad (10) \qquad Nu_c = \frac{h d_c}{k_f} \quad (11) \qquad h = \frac{q}{T_w - T_f} \quad (12)$$

The pressure loss (ΔP) and convective heat transfer coefficient (h,) are calculated after the flow is fully developed in the channel.

The boundary conditions are set as follows:

The velocity inlet is applied at the channel inlet: $u = u_{in}$, $v = w = 0$, $T = T_{in}$. The top and bottom (rough) walls are under constant heat flux, q =190 W/m^2. The remaining walls (side; smooth) are assumed adiabatic, $\partial T/\partial x= \partial T/\partial y=\partial T/\partial z=0$. For the outlet, the pressure is set, P_{out} = 0. The computational domain is shown in Fig. 1(a). The above mentioned governing equations and boundary conditions are solved by the finite volume method with Fluent program [12]. The first-order upwind scheme for discretization of governing equations along with the SIMPLE algorithm for the velocity-pressure coupling is utilized to perform the numerical analysis.

4. Results and discussion

4.1. Grid independence and model validation

Grid independence tests are conducted using several different mesh sizes, and criterions for discrepancies of f_c are applied. ANSYS-Workbench is adopted to generate mesh and roughness models. The channel width of 12.7 mm is considered for the independence criteria. The discrepancies of f_c between 9, 13 and 15 million are 3% and 1%, respectively, as discussed in Table 1. Thus, results from grid sensitivity analysis show that mesh composed by tetra grid with 13 million is suitable for the present study.

Table 1
Grid independence verification

#	Grid # (10^6)	f_c	Difference (%)
1	4.3	0.28	-
2	6.5	0.273	2.5
3	9	0.261	4.4
4	13	0.252	3.4
5	15	0.249	1.1

Table 2
Case study for model validation.

Roughness	Re_c	r (μm)	H (μm)	Nu_c (Present)	Nu_c [8]
Sinusoidal	100	30	250	10.67	11.33
Sinusoidal	100	50	250	8.58	8.601

The present model is applied in the cases from a reference study [8], to validate the model. The details are listed in Table 2, and deviations are in the acceptable range.

4.2. Fluid flow and heat transfer

In the present study, micro- and mini-channels structured with surface roughness are studied for varying aspect ratios (0.015, 0.021, 0.038) and respective widths (W) are 12.7 mm, 8.95 mm and 5 mm. Fig. 2(a) shows that constricted friction factor (f_c) is higher for lower aspect ratio of channel (r and H are constant), i.e. f_c of α_c = 0.015 are 5% and 9% higher than that of α_c = 0.021 and α_c = 0.038, respectively at Re_c = 100. It means that increasing surface area of rough walls increases the frictional effect and hydraulic diameter also increases with the width. These higher frictional effects are also due to the additional effects of sharp bends and sudden change of flow direction after every roughness curve (Fig. 3).

Fig. 2(b) shows that lowering the α_c results in the increase of convective heat transfer (Nu_c). It also elaborates that increasing Re_c improves the Nu_c for all the α_c variations. It is clear from Fig. 2 that wetted perimeter variation is dominating as compared to hydraulic diameter. The larger wetted perimeters mean more convective heat transfer, as explained by Newton`s law of cooling. A smaller hydraulic diameter of rectangular micro-channel has a larger wetted perimeter and convection heat transfer area (between the working fluid and heat sink). This increases the heat transfer from the heat sink to the cooling medium and leads to higher cooling performance of heat sinks. Hence, Nu_c for α_c = 0.015 is 8% and 20% higher than that of α_c = 0.021 and α_c = 0.038, respectively. Fig. 3 elaborates another factor which shows that the flow velocities are higher in the case of higher α_c (Fig. 3(a)), but the flow disturbances are more in the channels with lower α_c. This phenomenon seems additive reason of higher convective heat transfer. Although, the boundary layer thickness of stagnant flow seems same in both the cases.

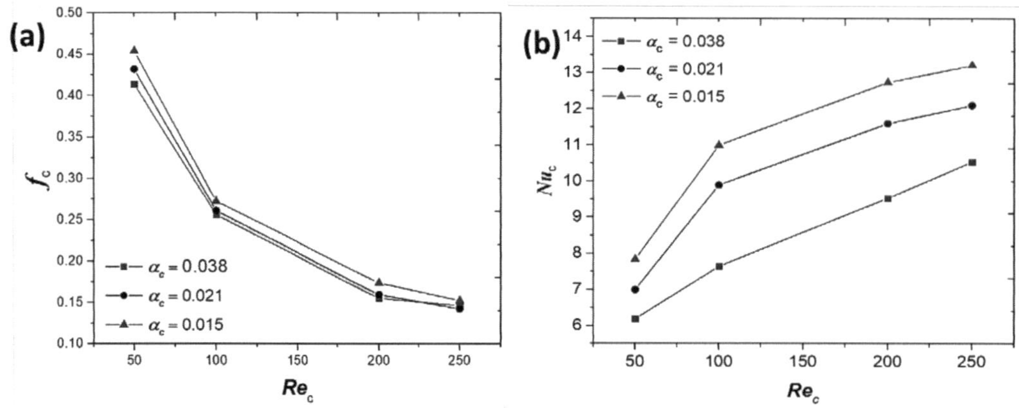

Fig. 2. Effect of channel aspect ratio on rough channels (a): f_c vs. Re_c; (b) Nu_c vs. Re_c.

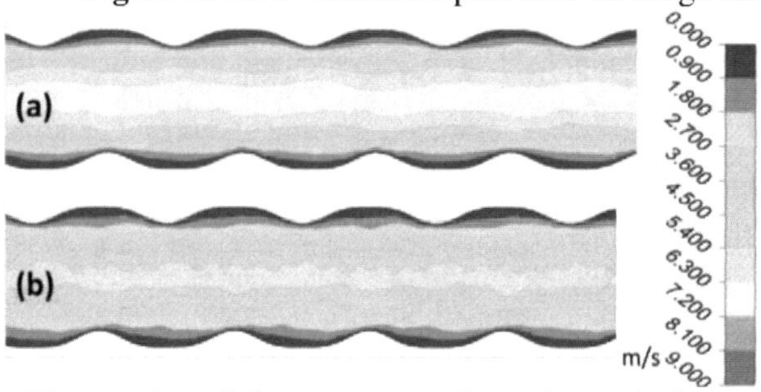

Fig. 3. Velocity contours of rough micro- and mini-channel heat sinks (Re_c = 100): (a) α_c = 0.038; (b) α_c = 0.021.

The results of the present study and correlations of smooth channels for Nusselt number are plotted in Fig. 4(a) to elaborate the comparison. The correlations are given below [1, 8],

$$Nu_s = 4.364 + \frac{0.0668(ReDPr/L)}{1+0.4(\frac{ReDPr}{L})^{2/3}} \tag{13}$$

$$Nu_{th} = 8.235(1 - 2.0421\alpha + 3.0853\alpha^2 - 2.4765\alpha^3 + 1.0578\alpha^4 - 0.1861\alpha^5) \tag{14}$$

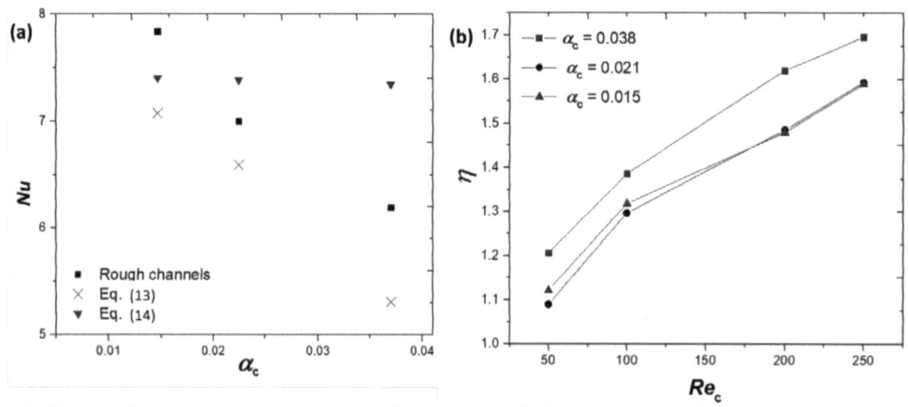

Fig. 4. (a) Comparison between present study and correlations, (Re_c = 100): Nu vs. α_c; (b) Effect of aspect ratio on rough channels: η vs. Re_c.

The calculated values (of Nu_c at Re_c = 100) are lower than the approximations of Eq. (14) (Shah and London) and higher than Eq. (13) (Grigull and Tartz). The trends of our simulations are close to the later one (Eq. (13)) because of the dependency on both Re_c and dimensional parameters of the channel. However, the higher heat exchange (Nu_c) of rough channels is the influence of structured sinusoidal roughness which leads to flow modifications and increased contact time with the heated wall. The α_c of these channels is very low as compared to size of the channels for which these correlations were developed and seems another reason of this discrepancy.

4.3. Thermohydraulic performance factor

The efficient heat sinks are the ones with maximum heat transfer and minimum frictional resistance. Thus, thermal performance factor η is gauged in terms of Nusselt number and frictional factor of both smooth and rough channels, given as

$$\eta = \frac{Nu_c/Nu_{th}}{(f_c/f_{th})^{1/3}} \tag{15}$$

where f_{th} and Nu_{th} represent the corresponding theoretical friction factor and Nusselt number for smooth channels, respectively. Eqs. (14) and (16) are applied for the calculation of respective values of Nusselt number and friction factor of smooth channels [8].

$$f_{th} = \frac{24}{Re}(1 - 1.3553\alpha + 1.9467\alpha^2 - 1.7012\alpha^3 + 0.9564\alpha^4 - 0.2537\alpha^5) \tag{16}$$

Fig. 4(b) shows that the thermal performance factor is more than 1 for all the considered cases due to the positive impact of surface roughness and increasing with increasing Re_c. The highest values are observed for the channel of $\alpha_c = 0.038$ due to the corresponding lower values of Nusselt number for smooth channels. The corresponding f_c is also lower for higher aspect ratio. The average rise of $\alpha_c = 0.038$ at $Re_c = 250$ is almost 6% more than the other two variations ($\alpha_c = 0.021$, $\alpha_c = 0.015$).

5. Conclusion

Numerical simulations are carried out to study the effect of aspect ratio (channel widths) on micro- and mini-channels structured with sinusoidal roughness, and conclusions are as follows:

1. The maximum deviation of the f_c among the variations is not more than 10%.
2. The constricted Nusselt number (Nu_c) depends on both constricted aspect ratio (α_c) and Reynolds number (Re_c). The values are also higher than those predicted by the correlations due to the presence of roughness structures.
3. The increase in channel width (decreasing aspect ratio) results in the escalation of convective heat transfer.
4. The performance factor is increasing with increasing Re_c and at higher α_c. All the values are higher than 1 with a maximum value of 1.65 (at $W = 5$ mm and $Re_c = 250$).

6. References

[1] M Asadi G Xie B Sunden 2014 England. Int. J. Heat Mass Trans. 79 34-53.
[2] W Qu I Mudawar 2002 England. Int. J. Heat Mass Trans. 45 2549-2565.
[3] X Yuan Z Tao H Li Y Tian 2016 China. Chinese Journal of Aeronautics, 29 1575-1581.
[4] R Dey T Das S Chakraborty 2012 United States. Heat Trans. Eng. 33 425-446.
[5] H Wang Z Chen J Gao 2016 England. Appl. Therm. Eng. 107 870-879.
[6] S Shen J Xu J Zhou Y Chen 2006 England. Energy Conv. Manag. 47 1311-1325.
[7] V Dharaiya S Kandlikar 2013 England. Int. J. Heat Mass Trans. 57 190-201.
[8] M Kharati-Koopaee M Zare 2015 France. Int. J. Therm. Sci. 90 9-23.
[9] C Zhang Y Chen M Shi 2010 Switzerland. Chem. Eng. Proc. 49 1188-1192.
[10] I Ghani N Kamaruzaman N Sidik 2017 England. Int. J. Heat Mass Trans. 108 1969-1981.
[11] L Guo H Xu L Gong 2015 England. Appl. Therm. Eng. 84 399-408.
[12] A. Fluent, Theory Guide and User's Guide 2015 USA. Ansys Inc.

Acknowledgment

This work is supported in part by Scientific Research Foundation for the Returned Overseas Chinese Scholars, State Education Ministry (No.2012940) and the Fundamental Research Funds for the Central Universities (09MG25).

IOP Conf. Series: Materials Science and Engineering **717** (2020) 012023 doi:10.1088/1757-899X/717/1/012023

Motion Analysis of Kick Mechanism Using in Muay Thai Matial Art

T Weerawat[1], S Jiraphon[2], Ch Kontorn[2] and Kh Sorada[2]

[1] Sports and Health Center, Suranaree University of Technology, 111 University Avenue, Suranaree Sub-District, Muang District, Nakhon Ratchasima 30000 Thailand
[2] Institute of Engineering, Suranaree University of Technology, 111 University Avenue, Suranaree Sub-District, Muang District, Nakhon Ratchasima 30000 Thailand

soradak@sut.ac.th

Abstract. The purposes of this research were to analyse speed of basic Muay Thai skills which consists of using feet and knee. By using the human motion-tracking cameras to create 3D animations that is associated with an image from a video camera and biomechanics analysis. The subjects group consisted of 5 professional boxers from Kiat Moo 9 Muay Thai camp by using purposive sampling method. The instrument used of this research was Qualisys Trach Managa (Motion Capture System). It was used to collect data motion capture and to analyses velocity Muay Thai basic skills. Statistics used for this research were arithmetic mean and standard deviation. The results showed that the average speed of basic Muay Thai skills include, Diagonal Kick, Straight Foot-thrust, Sideways Foot-thrust, Straight Knee-kick and Diagonal Knee-kick has 17.15, 6.01, 5.28 6.29 and 6.42 m/s, respectively. Standard deviation has 2.90, 1.47, 1.24, 1.01 and 1.75 respectively.

1. Introduction

Muay Thai, also known as Thai boxing, is a martial art that gradually developed over many centuries. Muay Thai was originally trained in to assert oneself in real fight situations. In the course of time Muay Thai also became popular as a competitive sport at fairs, which both gave athletes the opportunity to test their skills and provided entertainment for the audience. [1] Today, Muay Thai is practiced as a competitive sport, for fitness, and for self-defense. Muay Thai competitions are a spectacular sight and are extremely intensive. Fighters attack each other to full physical exhaustion with fists, elbows, kicking, and knee techniques. [2] Thai boxers generally have had excellent technical training, and their bodies have been trained to become muscular, high performance instruments.

However, Muay Thai training must understand and practice basic skills such as the use of foot, knee and elbow. [2] The basic training Muay Thai boxing to teachers to begin from the basic posture training is taking boxing stance positioning organs and the fist foot. Knee and elbow etc. and must practice regularly but the present teacher skilled boxer boxing training can be difficult.

For this reason, the application of sports science and sports science knowledge to the conservation and development of the fundamental Muay Thai is important so also must have a method to seek the facts or findings as the new information that will lead to the development of national sport even more. The studies and boxing skills by recording motion capture image. But it cannot be analyzed body

movements or the movement of the comparison with master trainers to use in training athletes in Thai boxing.
Therefore, the study and analysis of motion of a body's components in designing the basic skills in Thai boxing will help see the movement. [3-4] Each step carefully and clearly such as time, speed, acceleration and angle of movement [5-6], which is a component of the basic skills in Thai boxing will make people interested to get into the study and understanding of the movement. In addition, it can be used as a baseline in the design of teacher training student Thai boxing. Students and the general public, which is considered a conservation culture of Thailand.

2. Theory

2.1. Basic Muay Thai skill

In Muay Thai, leg work training is an important skill because the leg is a powerful organ for fighting and able to use at a distance. The leg work of Muay Thai defines from below the knee to the toes. The powerful stances of leg work are knee kick and the Foot trust. The Knee kick which the spinning body into the knee creating a devastating impact to the opponent, for example, the straight knee kick, the diagonal knee kick, etc. The foot trust which the method of bend at the knee, then quickly extending your leg to use foot or heel to attack the target, for example, Straight Foot thrust, Sideways Foot thrust, etc.

2.1.1. The Diagonal Kick. Figure 1 has shown the Diagonal kick post of the boxers. The boxer put his body-weight on his left leg and twisting his right hip then raise his right leg, bent at left. The lower leg is therefore at an approximate 45° angle, the shin and the toes being in a continuous straight line.

Finally, lash your lower right leg diagonally upwards from right to left against the target.

Figure 1. The diagonal kick. [7]

2.1.2. The Straight Knee – Kick. Lean forwards, transferring the body-weight to the left foot and firmly grasp the opponent's neck with both hands as shown in Figure 2.

With a twist of the right hip over left for extra force, thrust your right knee up against the target, the toes of your right foot pointing directly downwards to the floor, while simultaneously pulling down on your opponent's neck. (When adopting this method, your body will naturally lean slightly backwards, with the heel of your left foot raised off the ground).

2.1.3. The Diagonal Knee – Kick. Lean forwards, transferring the body-weight to the left foot and firmly grasp the opponent's neck with both hands.

Thrust your right knee diagonally upwards at an angle of approximately 45° from right to left, the toes of this foot straight down. Simultaneously, pull diagonally downwards from left to right on the opponent's neck, bracing your left foot against the floor as shown in Figure 3. Impact your knee-cap forcefully against the target.

Figure 2. The Straight Knee-Kick [7]

Figure 3. The Diagonal Knee-Kick [7]

2.1.4. The Straight Foot – Thrust. Put the body-weight on the left-leg and raise the right leg, the knee bent at a 90° angle, the toes pointing forwards as shown in Figure 4.
Twisting your right hip and thrusting your right leg straight.

a. your toes tightly pressed together
b. the ball of your foot, fully exposed by straining your toes
c. the sole of your foot
d. your heel to impact against the target

2.1.5. The Sideways Foot – Thrust. Pivot to the left on the left foot, while bringing the right hip round to the front (Figure 5). At the same time, pull your right leg, knee tightly bent, up and into your body, learning back and over to the left. Your right shoulder, hip and heel are thus in a straight diagonal line pointing to the target.
Thrust your right forwards so that the sole of the foot impact against the target.

Figure 4. The Straight Foot-Thrust [7]

Figure 5. The Sideway Foot-Thrust [7]

2.2. Motion Tracking Technology
Motion Tracking is important in research at present. Especially used to study and analyse the movement in athletic training. The motion tracking using the science of mechanics and techniques of using image such as [4] Motion-tracking cameras to study the movement of the human body by 3D simulation of motion of the arms and upper body, which can create an animation corresponding to the video. [3] In addition, modelling and tracking human motion by Function Analysis by building a statistical model for the movement of human activity. Consider that the human body consists of a

series of articulated cylinders and joint effects of tracking Algorithm can track the movement of the video as well. [4] For the presentation of a multiple hypothesis tracking which is classical approach by which the hypothesis is generated from the information that there is a relationship between a series of measurements and features target. [5]

Motion Tracking is important in research at present. Especially used to study and analyze the movement in athletic training. The motion tracking using the science of mechanics and techniques of using image such as [4] Motion-tracking cameras to study the movement of the human body by 3D simulation of motion of the arms and upper body, which can create an animation corresponding to the video. [3] In addition, modelling and tracking human motion by Function Analysis by building a statistical model for the movement of human activity. Consider that the human body consists of a series of articulated cylinders and joint effects of tracking Algorithm can track the movement of the video as well. [4] For the presentation of a multiple hypothesis tracking which is classical approach by which the hypothesis is generated from the information that there is a relationship between a series of measurements and features target. [5]

3. Methodology

In the research process to analyse the biomechanical basic skill Muay Thai by using Model-Based Approach [8-9]. The detailed research steps as follows.

Analysis of the movement of Muay Thai by using purposive sampling method of 5 professional boxers from Kiat Moo 9 Muaythai camp at Burirum province. The basic skills of Muay Thai for analysis which consists of Diagonal kick, Straight Foot-thrust, Sideways Foot-thrust, Straight Knee-kick and Diagonal Knee-kick. The instrument used of this research was Qualisys Track Manager (Motion Capture System) made in Sweden. It was used to collect data motion capture and analyze speed Muay Thai basic skills. The figure 6 and 7 show the Mark position on the boxer body on the front and back view and the detail of each mark appear in the Table 1. Statistics used for this research were arithmetic mean, standard deviation, maximum and minimum.

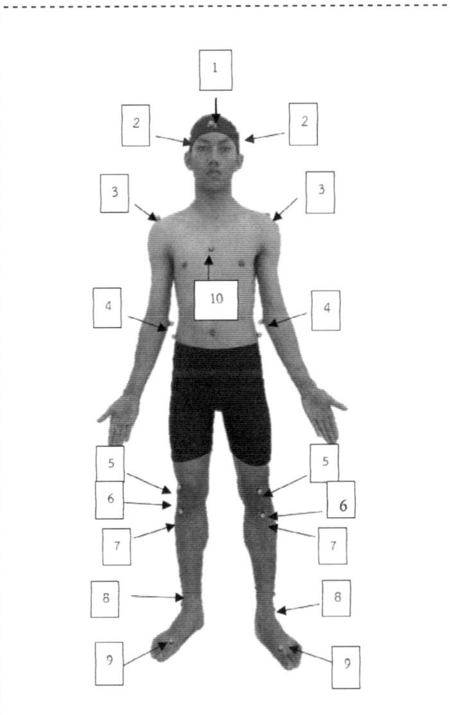

Figure 6. Mark position (Front) [7]

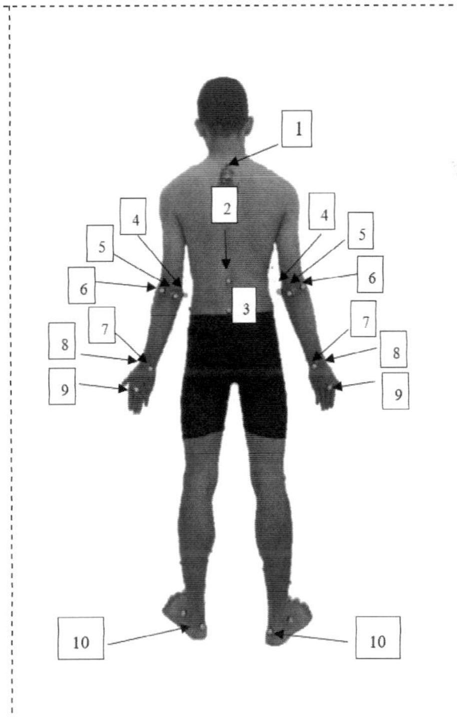

Figure 7. Mark position (Back) [7]

4. Results

Biomechanical Analysis of the Muay Thai skills of 5 professional boxers by considering the statistics of basic skills, including, Diagonal Kick (DK), Straight Knee-kick (SKK), Diagonal Knee-kick (DKK), Straight Foot-thrust (STFT) and Sideways Foot-thrust (SIFT) are shown in Table 2 and Figure 8-9.

Table 1. Details of the marker position (Front and Back).

	Abbreviation	Position (Front)	Abbreviation	Position (Back)
1	CFB	Center of Frontal Bone	C2	Cervical 2
2	RTPB, LTPB	Right and Left Parietal Bone	T12	Thoracic 12
3	RTAB, LTAB	Right and Left Acromion Bone	CSB	Center of Sacrum Bone
4	RTICB, LTLCB	Right and Left Iliac Crest Bone	RTCF, LTCF (Elbow)	Right and Left Coronoid Fossa
5	RTLCL, LTLCL	Right and Left Lateral Cartilage Ligament	RTME, LTME	Right and Left Medial Epicondyle
6	RTCPB, LTCPB	Right and Left Center of Patella Bone	RTLE, LTLE	Right and Left Lateral Epicondyle
7	RTHTB, LTHTB	Right and Left Head Tibia Bone	RTUB, LTUB(Insertion)	Right and Left Ulna Bone
8	RTTB, LTTB	Right and Left of Talus Bone	RTRB, LTRB (Insertion)	Right and Left Radial Bone
9	RTMTB, LTMTB	Right and Left Metatarsal Bone	RTLB, LTLB (Wrist)	Right and Left Lunate Bone
10	STB	Sternum Bone	RTCCB, LTCCB	Right and Left of Calcaneus Bone

Table 2. Statistical analysis of Muay Thai skill

Muay Thai skills	$\overline{X}$	SD	MAX	MIN
Diagonal Kick	17.15	2.90	21.03	13.58
Straight Knee-kick	6.29	1.01	7.95	6.18
Diagonal Knee-kick	6.42	1.75	8.68	3.87
Straight Foot-thrust	6.01	1.47	7.93	4.38
Sideways Foot-thrust	5.28	1.24	7.26	4.34

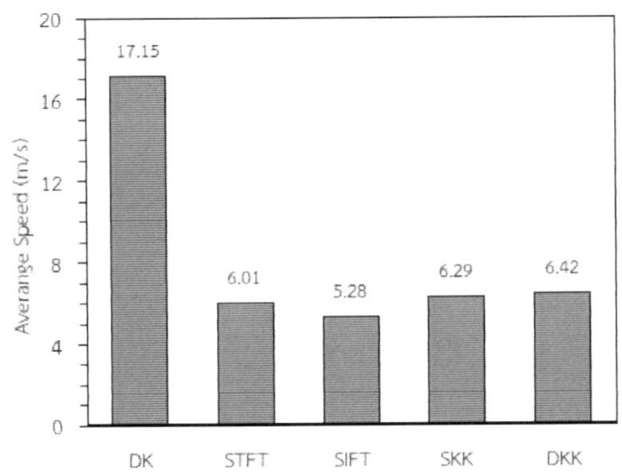

Figure 8. Average speed of Muay Thia Skills.

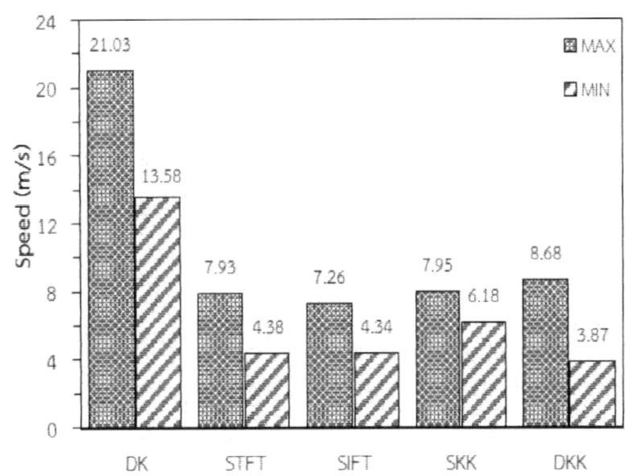

Figure 9. Maximum and minimum speed of Muay Thai skills

5. Conclusion

The Speed analysis by using Motion tracking technique of Muay Thai skills of 5 professional boxer reveals the target strength for novice athlete. The average speed of each Muay Thai trances in this study are 17.15, 6.29, 6.42, 6.01 and 5.28 m/s for Diagonal Kick, Straight Knee Kick, Diagonal Knee Kick, Straight Foot Kick and Sideways Foot Kick, respectively. The finding that Diagonal Kick could make the most impact to the opponent among legwork trances because of the highest speed. The future work of this study could explore the Thai boxing training program for the athlete by using speed data of the Muay Thai trances of professional boxers along with their physical fitness such as grip strength, flexibility, lung capacities, Leg strength, etc. The program expects to enhance the athlete's physical fitness close to the professional boxer. The biomechanical analysis of athletes using the Motion tracking technology could be applied to ensure the success of the program compares to the data of professional boxers.

6. References

[1] Saengsawang P, et al. 2540 Thai Boxing. Office of The National Culture Commission *Ministry of Education*

[2] Christoph D Muay Thai Training Exercises: The Ultimate Guide to Fitness, Strength and Fight Preparation. *Blue Snake Book*

[3] Kakadiaris I and Metaxas D 2000 Model-Based Estimation of 3D Human Motion *IEEE Transactions on Pattern Analysis and Machine Intelligence* Vol 2 No. 12 (Dec. 2000)

[4] Ormoneit D et al. 2000 Learning and Tracking Human Motion Using Function Analysis *IEEE Workshop on Human Modeling, Analysis and Synthesis* Hilton Head SC June 2000

[5] Cham T J and Rehg J A Multiple Hypothesis Approach to Figure Tracking *Proc.Conf.Computer Vision and Pattern Recognition* Vol 2 (June. 1999) pp 239-245

[6] Garrila D M 1999 The Virtual Analysis of Human Movement: A Survey *Computer Vision and Image Understanding* Vol 73 No 1 (Jan. 1999) pp 82-98

[7] Weerawat T 2017 Velocity analysis of fundamental Muay Thai skills *Doctoral Thesis* (Ratchaburi: Muban Chom Bueng Rajaphat University)

[8] Poppe R 2010 A survey on vision-based human action recognition *Image and Vision Computing* 28(2010) pp 976-990

[9] Weinland D, Ronfard R and Byer E 2011 A survey of vision-based methods for action representation, segmentation and recognition *Computer Vision and Image Understanding* 115(2011) pp 224-241

IOP Conf. Series: Materials Science and Engineering **717** (2020) 012023 doi:10.1088/1757-899X/717/1/012023

Acknowledgments
The authors would like to thank to Mr. Sutjai Pumprakhon who the boxing Manager, Boxers in Kiat Moo 9 Muay Thai Camp at Burirum Province and Suranaree University of Technology supported this research.

AUTHOR INDEX